Also from Westphalia Press
westphaliapress.org

POLICY AND ARMS

Policy and Arms

Westphalia Press
An imprint of Policy Studies Organization
1527 New Hampshire Ave., NW
Washington, D.C. 20036
info@ipsonet.org

ISBN-13: 978-1-63723-825-7

Cover design by Jeffrey Barnes:
jbarnesbook.design

Daniel Gutierrez-Sandoval, Executive Director
PSO and Westphalia Press

Updated material and comments on this edition
can be found at the Westphalia Press website:
www.westphaliapress.org

Policy and Arms

by Lieut-Colonel
Charles à Court Repington

WESTPHALIA PRESS
An imprint of
Policy Studies Organization

POLICY AND ARMS

By Lieut.-Colonel

CHARLES à COURT REPINGTON, p.s.c.

C.M.G.

Commander of the Order of Leopold,

Officer of the Legion of Honour.

AUTHOR OF

"THE WAR IN THE FAR EAST," 1905; "IMPERIAL STRATEGY," 1906;
"FOUNDATIONS OF REFORM," 1908; "ESSAYS AND CRITICISMS," 1911;
"VESTIGIA," 1919; "THE FIRST WORLD WAR," 1920;
"AFTER THE WAR," 1922.

LONDON: HUTCHINSON & CO.

PATERNOSTER ROW

1924

LIST OF MAPS

CONTENTS

CHAPTER XIII

THE CO-ORDINATION OF THE DEFENCE FORCES

CHAPTER XIV

IMPERIAL DEFENCE

CHAPTER XV

CHURCHILLIAN STRATEGY

CHAPTER XVI

SINGAPORE OR SYDNEY ?

CHAPTER XVII

THE N.W. FRONTIER OF INDIA

CHAPTER XVIII

AFGHANISTAN

CHAPTER XIX

FRONTIER TRAGEDIES

CHAPTER XX

INDIAN FRONTIER POLICY

APPENDICES

ALEXANDER GROSS F.R.G.S.

DISTRIBUTION OF THE UNITED STATES ARMY.

PREFACE

THIS volume contains chapters upon a number of subjects which belong to our peace and have engaged my attention during the last few years. On none of them have the public had a glimmer of light from the General Staff since the war ended.

I have begun with a sketch of the old diplomacy of the sixteenth century and have followed it up by a review of the present state of Europe before plunging into American affairs. The past and the present can thus be contrasted, and it will be seen that while the old diplomacy dealt mainly with only a few leading men and women, modern diplomacy has to deal with mass movements, mass opinions, and extends beyond the old confines of Western Europe to the whole of the world. But, expand though the stage may, and change though the players must, the best principles of the old diplomacy survive, and have only to be accommodated to new forces, new ideas, and new surroundings.

I have followed this by a glance at French ideas of war preparation, aerial policy, and the colonial army, and have also outlined Spain's military problem in Northern Morocco.

The questions of the Unified Command during the World War, and of the high command in future coalition armies, lead up to studies upon the co-ordination of our Defence Forces and Imperial Defence generally. Herein I have shown how we, almost alone among the Powers, stand disarmed in an armed world. What our public wishes, it gets ; but it cannot compel a critic to consider that the absence of serious military policy is reasonable at a moment when we are drawing near to, if we have not already entered upon, the most difficult period of our existence as an Empire.

A recent book by a glittering politician has compelled me to restate the main lines of the strategy of the Westerners during the World War, in order that the public may not be led astray by doctrinal heresies.

In the matter of Singapore as our Grand Fleet base in the Pacific, I have given my views which unfortunately differ from the decisions—I have heard no arguments—of our Home and Dominion authorities.

I am similarly unfortunate in having to oppose our present commitment in Waziristan. The N.W. Frontier of India is one of our greatest military problems. Therefore I have given a sketch of our past relations with Afghanistan, Persia, and the independent tribes, and have shown my objections to our present policy. It is my practice, when I criticise, to suggest an alternative policy in each case.

A military critic is a watch-dog of the public. He must form his own opinion of those who attempt to break into the house of strategic principle and sound tradition, and must give warning to his masters. They can move, or not, as they please. The watch-dog's duty is done.

My grateful thanks are due to Viscount Burnham, proprietor of the *Daily Telegraph;* to Mr. George Blackwood, of *Blackwood's Magazine;* to Mr. George Dewar, Editor of the *Nineteenth Century and After;* and to the Editor of the *Atlantic Monthly* for allowing me to incorporate in this book articles which have appeared in their pages.

POLICY AND ARMS

CHAPTER I

The Old Diplomacy: (I) Elizabeth

The Art of Diplomacy—The Elizabethan Period—Diplomatic Situation—Dangers Besetting Protestant England—Character of Elizabeth—Perils of her Youth—Education—Characteristics—Lord Burghley—Character and Qualities—Internal Reforms—Burghley's Maxims—Methods and Manners of the Times—Secret Service—Walsingham—Foreign Policy—Elizabeth and Scotland—The Second Marriage of the Queen of Scots—Proceedings of the Privy Council—The Negotiations for a Treaty with France—D'Alençon and Elizabeth's Favourites—Elizabeth, a Friend to France—The Privy Council's Methods of Work—England's Sober Statesmanship Preserves Peace.

We hear a great deal about the old diplomacy in these days, and usually hear it compared most favourably with modern practice. Men grown grey in the diplomatic service bewail the decadence of their profession, which, they say, lacks information, is not adequately consulted, is not made the real medium for international intercourse, is elbowed aside, and finds its *rôle* assumed by Missions and Conferences, Councils and Leagues, which leave this poor suffering world considerably worse off than they found it. Now that the unregretted demise of the Coalition has restored our Foreign Office to its proper sphere of activities it is convenient to throw a glance back upon the past, and to observe how our foreign relations were conducted by recognised masters of the craft.

In order to be able to appreciate what the old diplomacy was we must begin by being almost offensively industrious. We must read the lives and memoirs of the old diplomatists, and must have a good knowledge of their times and of the politics and personalities of the period. This involves, for an Englishman, an examination of the work of Rosebery and Salisbury, of Clarendon and Palmerston, of Wellesley and Canning, of Castlereagh and the Pitts, and then when he has read the memoirs of the contemporary ambassadors he must continue with Walpole, Burghley, and Wolsey, not to speak of others, to the very threshold of the Middle Ages. The history of the foreign nations with whom we were dealing at the time, the lives of their great men and women, and the foreign point of view are no less important, because we have regretfully to observe that almost all history is written from a national point of view and with a national, not to say a bigoted, bias ; so it is necessary to observe the period in question with most judicious eyes, and to separate permanent and doctrinal diplomatic practices from those which are temporary and evanescent.

We desire in each case to know how the great masters and artists in foreign policy were prepared by education, training, and experience for the mission which they successfully fulfilled, on what lines they worked, how they acquired their information, how they chose their agents, how they organised their offices, how they sent and received reports, how they influenced the Courts and Councils of their time, what general doctrines they followed or created, and last, but not least, what part of their doctrines we can carry forward as valuable for the guidance of diplomacy in the future.

It is not correct to suppose that the condition of the modern world is so different from the past that research is wasted. Diplomacy is the art of dealing with men, women, and things for the profit of one's own nation. If to-day, by reason of the spread of democratic ideals, we have to add that it is also for the profit of the world, we can observe that though this admirable addendum is the source of much rhapsody and rhetoric, and the text for many resounding perorations, the modern diplomatist does not forget his nation's interests, except by inadvertence, and, in fact, is guided by them just as much as his predecessor in the past. Men and women remain as of yore complicated creatures, by no means immune from the preferences, prejudices, and passions of ancient days. In a word, diplomacy is the art of dealing with men, and men have not changed much since the dawn of history.

The Elizabethan Period

The sixteenth century was the century of the Reformation. Luther died in 1546, Calvin in 1564. From the Reformation came modern England, and from the Puritan element of the Reformation there came modern America. It was a great period, and a time when the renascence of great ideas in art and letters spread over the States of Western Europe. Machiavelli belonged to the century, and had written *The Prince* before Elizabeth came to the throne, tearing the veil from the hypocrisy of diplomacy and thereby incurring the malediction of the powers of the day, for nobody, of course, likes to be shown up. For England it was the century of the Tudors, matchless in governance and statecraft; for Scotland, the century of the Stuarts and the Congregation, who were quite the reverse; for Ireland, a century like others, with rebellion, repression, crime, and counter-crime following their customary and mournful course.

The second half of the century was the Elizabethan period, fruitful in wise men, great women, and effulgent ideas; with a galaxy of talent round the throne unequalled in merit either before or since. Throughout the half-century Queen Elizabeth reigned, and her great Minister William Cecil, Lord Burghley, piloted the ship of State. To assign the precise share of credit which belongs to Queen and Minister for the successful conduct of foreign affairs throughout this period is a difficult task, so inextricably are the characters of the two interwoven together up to the day of

Burghley's death. As a broad generalisation we can say that Cecil advised and Elizabeth decided. But it would be more correct to say that thanks to the co-operation of these two choice spirits, and to the aid of many other loyal servitors, at a time when loyalty to the Crown was still imperfect, England came through great dangers unscathed, almost without waging serious war, and, from being a little country surrounded with powerful enemies, grew into a Great Power, shaped and disposed to fulfil those high destinies which have since been accomplished, not only on both sides of the Atlantic, but throughout the world.

England was neither a powerful nor a united country when Elizabeth ascended the throne. A little population of three million souls in the days of Henry VIII., and possibly four millions in the time of Elizabeth, was faced by some sixteen millions of Frenchmen and about the same number of people under Philip II. of Spain. England became officially Protestant after Mary Tudor's death, but was divided on religious questions throughout the new reign. The process of conversion from the old faith to Protestantism was never quite completed, and in its first stages was slow, prudent, and unhasting. The majority of the Scottish nobility and landed gentry were still Catholic. So were many of the English, especially in the North and West, and not a few placed their duty to Rome before that to the Queen. The South of England, the Midlands, and the towns, London and Edinburgh, were the Protestant strongholds, and the Queen's progresses were almost wholly restricted to the home counties.

Rome was the implacable foe of Elizabeth from first to last, even if the character of the hostility varied with that of successive Popes. Ireland was Catholic, and Marie Stuart, throughout her troubled life, was devoted to the same faith. A combination of Rome with the Catholic Powers against Elizabeth, and the participation by Scotland, Ireland, and the Catholics of England in the coalition, was always a possibility, and often a dreaded and imminent possibility, throughout Elizabeth's reign. The excommunication of Elizabeth by the Pope in 1570 absolved good Catholics from their allegiance to her in the eyes of the Roman Church, and thereafter came the Ridolfi, Babington, and other plots, which aimed not only at the forcible reconversion of England to Catholicism, but at the Queen's life and at the lives of the chief members of her Council. The claim openly made by Marie Stuart to the throne of England and Ireland when she married the Dauphin of France, even if engineered by the French King, was an additional thorn in Elizabeth's side, and the more because many people in England considered Marie's claims to the throne better than those of the Queen.

These dangers created many and serious difficulties for Elizabeth and her Council. It was necessary for them to pass from the old religion to the new by slow stages, and with the least possible interference with the customs of the people. Ireland needed firm government, while in Scotland Elizabeth supported the Protestants and helped them, by a well-timed stroke of diplomacy and arms, to evict the French troops which had been sent to the support of the Regent Queen-Mother. On the continent of Europe the policy was, above all else, to prevent a combination of the Catholic Powers against England, and surreptitiously to support the Protestants in Germany, the Low Countries, and France, and in fact to resist Catholic ascendancy everywhere without involving England in a regular war. It was a policy of balance and compromise requiring extraordinary sagacity and finesse. England's friendship and alliances might change, but the objects of the policy never changed. The hand of Elizabeth was sought by all the marriageable princes of Europe, and courtship became part of the great game of diplomacy, and a part in which Elizabeth was peerless. She kept the princes dancing attendance as long as statecraft required, thereby influencing the action of their countries to her own advantage, but she never committed herself to an irrevocable step.

At the same time the Queen and her Council reformed the

currency at home, aided the merchants, and sought new outlets for their trade, while the Royal and Mercantile Navies were increased, and an indulgent eye was turned upon a number of ventures oversea, against Spanish commerce in particular, which partook of a piratical character and brought Spain into war at long last, but meanwhile trained the country to sea affairs and rendered the English seamen capable of affronting the dominant navy of the day with confidence in success. The great safeguard of the times, slowly though the fact dawned even upon Cecil, was that Spain could not afford to allow France to dominate in the Spanish Low Countries or in England or Scotland, while France desired to abate the power of Spain and would not follow her when she aimed at England. If the Low Countries figured so prominently in the events of the time it was largely due to the fact that they were the best market for English merchandise and the door through which the products of Europe best reached England. Not many doors for trade were open then, for the Turks controlled the Eastern Mediterranean, Spain had forestalled England across the Atlantic, and the Muscovy Company and other similar associations were affairs of recent growth.

Character of Elizabeth

When the Lady Elizabeth ascended the throne in 1558, and Cecil began his forty years of faithful service to her, both Queen and Secretary were well prepared by hard experience to guide policy and to control themselves. From the moment when Anne Boleyn was executed until the day when the courtiers trooped to Elizabeth's retreat at Hatfield to worship the rising sun her lot was one of sorrows, difficulties, and dangers. She passed through the hands of stepmothers who were not unkind to her, but whose over-rapid succession must have implanted in her mind a distrust of matrimony. She refused the hand of the Protector's brother when she was fifteen, and only her clear head and admirable courage saved her from falling with the Seymours. Accused later of complicity with the Wyatt rebellion and of intriguing with Edward Courtney, she was found to have done nothing to compromise herself, but she was consigned to imprisonment in the Tower, and if her enemies could have had their way she would have been brought to the block. Though outwardly conforming with the old religion in Mary Tudor's reign, she was the recognised head of the Protestants in England, and remained in danger until her sister's death. While Marie Stuart spent her youth surrounded by the luxury and splendour of the French Court, her life-long rival lived her first twenty-five years in a hard school of adversity, never perfectly sure what the next day might bring forth. All the time her coolness, courage, and self-restraint never left her, but it was a bitter experience for a girl fond of gaiety and amusement. Religiously minded Elizabeth was not. She would have liked an Anglican Church with Romish embellishments. There was nothing of the Puritan about her. She had very little liking for John Knox and disapproved of his famous *First Blast of the Trumpet against the Monstruous Regiment* [*rule*] *of Women* as much as Marie Stuart or Catherine de Médicis herself.

But during all these years of apprenticeship to sovereignty Elizabeth's education was not neglected, and she was in the first flight of the learned ladies of her day. Her father had made a great point that his children should be highly educated by the best English and foreign masters and mistresses, and the retired life led by the Princess until her accession was favourable for her studies. Henry VIII. had been well educated. Mo mean author, honoured by Rome for his theological disquisitions, and well schooled in Latin and modern languages, he chose for his children the best talent of the colleges. John Cheke, William Grindal, and Roger Ascham were the most competent tutors and the best scholars of the day, while Lady Tyrwhitt was an admirable governess. The pupil was certainly an apt scholar. She had immense application, a retentive memory, and almost a man's mind. She soon knew Latin and Greek well, besides several modern languages, and continued her readings and her study of them after her accession. Her crushing impromptu reply in Latin to a Polish ambassador who had been insolent provoked the amused admiration of the Court.

It can occasion no surprise that Latin was the *lingua franca* of the times, the medium preferred by historians like Camden, Scottish scholars like George Buchanan, and statesmen like Wolsey and Cecil. English had not then acquired the volume and flexibility of to-day, and as a medium for conveying precise thought Latin was supreme. The language of France differed even more from modern practice, and as for Scots letters, not even the penetrating and acute remarks of Maitland of Lethington inspire in us the wish to read many of them. Latin was the language of treaties as it was of the Mass. Cecil's private diary was written in Latin in abbreviated form, and even Latin numerals were used in his accounts. The letters of great classical scholar-statesmen like Sir Thomas Smith—and very amusing letters they often were—are interlarded with Latin and apt quotations. So, not uncommonly, are the Queen's.

The tributes to Elizabeth's learning are many, and most of them ring true. The compliments paid to her sagacity and statecraft by the far from friendly foreign ambassadors at her Court (Feria, Guzman, Aquila, Mendoza, and the rest), in their secret reports, testify to her superiority, even when one of them suggests that she has a hundred thousand devils in her, and another that she must be the daughter of the devil. She could outplay them all at their own game and they did not like it. Her clear and simple style of writing, her superb handwriting and gorgeous signature, fascinate us to this day.

In her wisdom, not to speak of her thrift, Elizabeth recalled her sagacious grandfather Henry VII.; in her occasional outbursts of furious passion and in her love of display she was her father's daughter ; while her lighter moods were reminiscent of poor Anne Boleyn. She had many faults. She was slow in reaching resolutions. In her later days she was barbaric in her dress. She was not very delicate in her language when angry, and her oaths are traditional. She was an accomplished coquette, and cared not at all that her blatant and most public love-making with a favourite, a suitor, or a foreign courtier should amaze the Court and cause scandalous tales of her behaviour to be spread throughout Europe. She was the embodiment of vanity, and the carefully prepared *mise-en-scène* when an envoy was adroitly placed to surprise her dancing, playing, or singing must have been more comical than convincing. She was an adept with the cross-bow, and often brought down half a dozen does during the very unsportsmanlike drives in the parks of her nobles.

There was a vein of frivolity running through her to which Leicester above all pandered. But they had been companions, and almost brother and sister, since the age of eight, and there was a touch of the schoolgirl about it all while youth still excused it, flashing out into the temper of a hoyden in later years, when she soundly boxed Essex's ears, tore a foreign order from a courtier's coat, or sent her high-heeled shoe flying across the room with unerring aim at Walsingham's head. Her rage became ungovernable when one of her favourites—Leicester, Essex, or Raleigh—secretly married, and inferentially the Queen's favourite of the moment was not authorised to look elsewhere. She was incredibly mean to Davison in the matter of the warrant for the execution of the Scottish Queen if we accept his story of the occurrences, and to many of her faithful councillors she was stinginess itself.

But if Cecil could afford to wink one of those great luminous eyes of his in benevolent indulgence at the Queen's foibles, then surely so can we, recalling the many great merits of her character and her statesmanship. She brought with her to the throne a thorough education, a great knowledge of the world, profound judgment of men, most of whom she could see through, and an equipoise of mind combined with unfailing prudence and caution in State affairs. If she hated to spend money, the Treasury benefited, but to her favourites and foreign visitors she was open-handed. Her generosity was returned by her nobles when they entertained her during her progresses.

Her twelve visits to Burleigh cost the Lord Treasurer 2,000*l.* to 3,000*l.* on each occasion, which we must multiply many times to compare with the value of money to-day. She certainly gave royally, but she expected her favourites to entertain on the same scale, and many of them ruined themselves in the process, or in supplying ships and troops for national purposes when Elizabeth refused to find them. She made her Court the centre of attraction for all the youth and chivalry of the day, and her judgment of character and quickness of eye enabled her to pick out the best and advance them. She had her father's quick eye for a man and her own discernment for a good-looking one. She rarely dismissed a councillor or a courtier, and her kindness to Cecil when he grew old greatly honoured her.

She bore herself with unequalled majesty in great State ceremonials, and men went in fear of her ; but among her people she was a popular favourite and conversed with them in such familiar fashion that she won all hearts and deserved her name of Good Queen Bess. She was a great character and a great Queen, and brought to her Council the priceless support of an almost unerring sagacity.

In the first twelve years of her reign she advanced but four men to the rank of Baron. She knew her own mind and made all her marriage projects subservient to State interests, while determining in her own mind, and repeating since the age of fifteen, that she would never marry. Accept this statement as a literal truth, and Elizabeth's love affairs are explained. But all the fervent love-letters to her swains are certainly not excused. In love-making no one should ever have credited a word she said.

But her courage was great, and most conspicuous in the hours of greatest danger. She was every inch a Queen.

Lord Burghley

William Cecil was no tyro in statecraft when, at the age of thirty-eight, he became the Queen's first Secretary of State. At Stamford and Grantham, and later at St. John's College, Cambridge, he had worked hard and had displayed excellent abilities which were subsequently confirmed during his studies at Gray's Inn, which were followed by his appointment to an office in the Court of Common Pleas. Taken up by the Protector Somerset during the reign of Edward VI., Cecil accompanied his patron to Scotland, and was present at the battle of Pinkie. He sat in Parliament in 1547, became Somerset's State Secretary, and was imprisoned at his fall, but was soon released and returned to office in 1550 as Secretary of State and member of the Privy Council, making himself as indispensable to Northumberland as he had been to Somerset, and remaining similarly in the background during the trial of his new chief. In 1554 and 1555 he was employed on diplomatic missions abroad in positions of no great prominence, but the experience was useful to him. He sat for Lincoln in the Parliament of 1555, but not in that of 1558, for he was now in close touch with Elizabeth, and was preparing for the death of Mary Tudor, which was foreseen.

When the event occurred it was found that Cecil had drawn up a document providing in all detail for Elizabeth's accession, that her proclamation was already drafted by him and that he had assumed the direction of her government in his old post. He retained the position of principal Secretary of State till he became Lord Treasurer, but the most important foreign affairs passed through his hands until his death in 1598. His barony was conferred in 1571, the fourteenth year of the Queen's reign.

Never did Elizabeth display more wisdom than when she

appointed Cecil to his post, and in no way did she better display the constancy of her character than by supporting him throughout his official life as she did. It was not as though the age was poor in statecraft or in character. As we meditate over the names of the Elizabethan nobility and gentry, we can easily suggest some half a hundred men who were worthy of places in her Council, men whom we would be proud, not to say relieved, to see now in a British Cabinet. A Council which could not find room for such a wise head as Walter Raleigh's must have been rich in talent indeed. Yet Cecil, we recognise, was unique. Not a finished courtier and still less a gallant, hating war and unskilled in arms, careless of exercise and ignorant of games, short in stature and sober in his dress, he was essentially the statesman with civilian mind. Work was his amusement, his library was his joy, and his gardens his relaxation. His capacity for work can only be described as prodigious. He was the factotum of the Government, and nothing escaped his observation or was too small for his attention. Religion, judicature, taxes and subsidies, customs, commerce, ships and soldiers, as well as his own considerable affairs, to which he paid meticulous attention, as he did to those of every one of the petitioners who wrote to him, seem to us enough to have exhausted the energies and overloaded the time of the most active of men. Yet he remained in close attendance on the Queen or the Council, and gradually concentrated in his hands all the strings of foreign affairs until he was completely master of them.

Cecil had many enemies. There are episodes in his early career which show that he had no instinct of self-sacrifice, nor any hankering for martyrdom. The fall of a patron was to him an episode which he allowed to pass by as a mere incident of history. But he was no political schemer. Dignified, sedate, balanced and most moderate, he thought out his counsel, delivered it, however unpalatable to others, and if the opposition was too strong he had a positive genius for stage exits and diplomatic illnesses. With all the knowledge in his wise head he was indispensable, and when his colleagues found themselves in difficulties they had to beseech him to return and get them out. He came back without reproaches and without resentment, coolly took up the argument where he had left it, and by sheer force of character and acquired knowledge carried conviction to all minds. He was very calm and very judicial. He was seldom deflected from a course, and though he despised heat in argument, and would never give tit for tat in debate, by a gentle obstinacy he returned to his point and had his way. To the fury of ambassadors he returned the soft answers that turn away wrath, and his very moderation was a force.

The relations between the Queen and Cecil have been so well described by themselves that the following letters may be quoted :

I am no opiniaster but an opynor [he wrote to his son Robert]. As long as I may be allowed to give advice [he wrote in 1595] I will not change my opinion by affirming the contrary, for that were to offend God, to whom I am sworn first, but as a servant of God I will obey Her Majesty's commandment and in no wise contrary the same, presuming that she being God's chief Minister here, it shall be God's will to have her commandment obeyed, after that I have performed my duty as a Councillor, and shall in my heart wish her commandments to have such good successes, as I am sure she intendeth. You see I am a mixture of divinity and policy, preferring in policy Her Majesty afore all others on the earth, and in divinity the King of heaven above all betwixt Alpha and Omega.

Serve God by serving the Queen [he wrote in his last letter to the same son], for all other service is indeed bondage to the devil.

Sir Spirit [wrote the Queen to Cecil one day], I doubt I do nickname you, for those of your kind (they say) have no sense, but I have of late seen an *ecce signum*, that if an ass kick you, you feel it too soon. I will recant you from being my spirit, if ever I perceive that you disdain not such a feeling. Serve God, fear the King, and be a good fellow to the rest. Let never care appear in you for such a rumour, but let them well know that you desire the righting of such wrongs, by making known the error, then you be so silly a soul as to foreslowe that you ought to do, or not freely deliver what you think meetest, and pass of no man so much, as not to regard her trust, who puts it in you. God bless you and long may you last. *Omnino.*—E. R.

The trouble that we experience in tracing the conduct of great affairs in Elizabeth's day is partly due to the bias of historians who allow their religious preferences and prejudices to colour their stories, and partly to the secretiveness and double-dealing of the times when language was often used to conceal thought rather than to express it. But we see clearly enough, especially since the Spanish manuscripts have been laid bare, that the conduct of foreign policy required sleepless vigilance. Cecil followed up the policy of Henry VIII. and Wolsey in his endeavour to hold the balance between France and Spain, to prevent the union of these two Catholic States against England, and to weaken them by aiding their rebellious subjects unofficially, so that England, while gaining her point, might not be committed to open hostilities. Cecil detested war, and so did the Queen, largely owing to its cost. Walsingham would not advocate wars because he considered their issue doubtful, and it must be admitted that the pursuit of Continental policy in arms brought to the reign very little profit. It deserved no more, for it was usually conducted in an underhand way and on every sort of pretext except the real one.

The Council often appears timid and irresolute in action, both at home and abroad, until the plots against the Queen's life goaded them into reprisals, but we have to remember that a capital defect of government in those days was the absence of police and soldiers, which left the field only too open to intrigue, while the cost of war revolted the Queen's economical mind.

Internal Reforms

But this penuriousness of the Queen which has been so frequently blamed was to a great extent the cause of her popularity. She must have had a clear recollection that the exactions of her father and his great Minister had produced the contrary result, with all the impediments that it then placed in the way of the prosecution of a strong foreign policy. At Elizabeth's accession her Treasury was at a low ebb and the coinage debased; her armouries were depleted; arms and weapons, as well as powder, could only be obtained abroad; and the Navy was ill-appointed.

She was at great charges on many occasions—to hold the Scottish border, to maintain her troops in Ireland and in the Low Countries, and to finance those princes who were her friends. In such conditions economy was a necessity for her, and she found that there was no better means of making her people contented than the reductions of their burdens to the lowest terms.

In this branch Burghley excelled. 'He looked narrowly,' says the admirable Camden, 'unto those who had the charge of Customs and Imposts by whose Avarice many things were underhand imbezeled and through whose Negligence the just Dues were not exacted.' The person who farmed the customs on payment of 14,000*l.* a year was made to pay 42,000*l.* and to hand over other sums for profiteering in the past. It was not permitted that the Treasury should be filled by the spoiling of the poorer people. Extortions and rigid contributions were banned, and Elizabeth would not suffer the poll-tax to be so much as mentioned. Thus released from the rigours of the Wolsey system, the people granted subsidies cheerfully to a larger amount than the Estates of the Realm estimated them. It was the policy to make as few demands upon the Estates as possible, for it was known that taxes were the greatest detriment to popularity, and that the poverty of a people was a prime cause of revolt.

If conflicts with the Commons, in Elizabeth's day, occupy but a small place in the history of the reign, we must ascribe it to the frugality of Cecil and to his regard for the revenue, which was ever a royal road to the heart of the English Estates. The Commons met rarely, and seldom sat for long. What most concerned them was the unsettled succession and the precariousness of the Queen's life. For this reason they attempted in 1562, 1566, and 1575 to force Elizabeth to marry or name a successor, but met their match in the Queen, who would never take either course, though she was prodigal enough in soothing words.

The Estates were in fact silenced by the frugality of the Government. By remitting a subsidy already voted she pacified the Commons in 1566. Cecil often resorted to the ancient practice of sending privy seals to borrow money from the rich, but they were repaid punctually, and their loans were only anticipations of the regular revenue. Drake and his comrades often poured wealth into Elizabeth's lap after some successful raids on Spanish territories and commerce, but in such adventures Cecil was never a partaker nor a shareholder. While Philip was ruining his country with the *alcabala* owing to his perpetual wars, England grew rich by legitimate trade, economical administration, and the avoidance of war, while the Commons, mainly Protestant with a strong Puritan element, had no quarrel with such a provident administration, and were ready for any sacrifices in the cause of the new religion.

In a few notable maxims Cecil has handed down to us the main features of his economic policy. 'The realm cannot be rich whose coin is base'—a maxim carried out in practice by the reform of the currency, which was one of the most popular measures of the reign. 'A realm can never be rich that hath not an intercourse and trade of merchandise with other nations'—and this trade he sought to extend, and extended, by every means open to statesmanship. 'A realm must needs be poor that carryeth not out more merchandise than it bringeth in'—the 'favourable balance of trade' principle in a simpler form, and applied by a succession of measures at home and abroad, to fisheries as well as to land commerce. 'War is a curse, and peace the blessing of God upon a nation,' was proved by his own experience, as was his other aphorism 'that a realm gaineth more by one year's peace than by ten years' war,' which truth the present generation, at least, has no need to learn. Wars other than the defensive war of the Armada year, and the surreptitious expeditions to help Huguenots or Dutch, Elizabeth never waged. She was the first ruler of England to stay at home, and neither wars of aggression nor annexations ever entered her mind or Cecil's. This successful home policy was the foundation of a sound foreign policy, for such policy based on an over-taxed and discontented people must fail.

Methods and Manners

Though the aristocratic habit of the 'grand tour' belongs to a later period, a large number of the gilded youth of the Elizabethan age went abroad to perfect themselves in the languages and literature of Europe. Some, no doubt, profited as little as Cecil's eldest son, but others brought back varied and useful knowledge, and gave to the Court that peculiar polish and refinement which was not the least of its charms.

The country became rich in men competent to undertake foreign missions, and to supply not only resident ambassadors at foreign Courts, but also those special missions which were so frequently despatched, whether for reasons of Court etiquette or of State. The ambassadors of England to France and Scotland, and in a lesser degree to Spain and the Emperor, were the main agents of information to Cecil and Walsingham in their dealings with foreign affairs. Men like Throgmorton in Paris and Randolph in Edinburgh were extremely well posted, and their despatches and letters from the capitals and Courts to which they were accredited were very valuable to Cecil. They did not stand alone, for Fénélon and Castelnau for France, and the various Spanish ambassadors, wrote lucid and entertaining reports home to their Governments, while Du Croc's reports from Scotland display accurate knowledge of events. The visits to England of Scottish Protestants like Murray, Morton, and Maitland, and of Melville and other emissaries of Marie Stuart, illuminated the darkness of the North.

There was no Press in those days and books were severely censored, though occasionally some diatribe or scurrilous publication obtained a wide circulation notwithstanding. Therefore these despatches and the reports of travellers were the main sources of information, and they were often rendered more valuable by the bribery that was so generally used in order to secure important news. It is probable that Rome, Paris, and Madrid were better informed by corrupt means than London. Many times we read of boasts by ambassadors to England that nothing escaped them whether in the Court or Council, and they were often so well informed that we can almost credit the rumours that they had spies in the highest circles. The faithful servants of Marie Stuart, after her flight to England, also carried news, projects, and intrigues far and wide, until it became evident at last that something had to be done to stop a practice which involved danger to the State, the Council, and the Queen.

Cecil was a man who kept his own counsel. His disposition was conciliatory and he was modest and retiring. He never attempted, nor thought of attempting, to exalt himself. There was no need in his days to curry favour with the mob. The support of the Crown sufficed. But he was the embodiment of knowledge and vigilance, and when the underhand methods of England's enemies were laid bare he set Walsingham to work to counteract them. Walsingham must be regarded as the father of English secret service. Before long he had covered Europe with his agents, and it was thanks to his efficient system that the most serious plots against the Queen were unravelled, the plotters seized and executed, and Marie Stuart finally condemned to death. The usual method was to insert an agent into circles whence danger was apprehended, to discover what was in the wind, to seize, steal, or buy the letters passing, to open them, copy them, and decipher them, and then to close them up so that no sign of tampering could be seen. There were experts in all these branches, and no cipher was safe. When all was in readiness, the blow fell and all the plotters were arrested. The culprit—and occasionally an ambassador—was dumbfounded by being confronted with his own letters and translations of his ciphers, and for minor culprits the rack or promises of rewards filled up the gaps of evidence in the case. The occasion on which a Spanish ambassador was shown up, commanded to attend the Council, and solemnly dismissed must have caused a great sensation at the time.

But there were occasions—and Marie Stuart's case is the most glaring—in which the procedure was wholly indefensible, and it is probable that a modern Court of Justice would not have condemned her. That she plotted against England with any and every foreign and domestic enemy of Elizabeth ; that it was a question of her life or Elizabeth's ; and that Elizabeth stood at last alone among her subjects unwilling to condemn, does not alter the fact that the case for the prosecution, and the methods employed, fill us with disgust, as does the rank injustice of the trial at Fotheringay itself. In keeping with the times these practices may have been, but they were a blemish upon the reign, and Elizabeth knew it.

But, if we place secret service aside, the foreign policy of the day was meritorious, and the manner in which Cecil prepared his cases and presented them to the Queen or the Council was a model of judicious procedure. To explain and to trace his methods it is best to take two concrete cases, one of internal and the other of external policy, and we may select the cases of the marriage of Marie Stuart as our first model and that of the projected alliance with France in 1581 as the second.

Elizabeth and Scotland

Throughout Elizabeth's reign Scotland was a source of danger. Unable to resist unaided the might of England, despite the glorious valour of her fierce and hardy mountaineers, Scotland looked abroad for succour, and had not far to look. France, with her customary political acumen, knew that a strong Scotland would prevent England from deploying her whole might against France, and therefore posed as Scotland's friend, assisted her with money and men, and left no stone unturned to injure England through her northern neighbour. It was for this reason that

Marie Stuart was transferred to France, confirmed in the Roman Catholic Faith by her astute uncle, married to the Dauphin while he was still almost a schoolboy, and made to claim the title of Queen of England and Ireland and to use the English Royal Arms. It was a foolish, empty, and highly dangerous piece of bravado. The injury touched Elizabeth in her most sensitive spot, that is to say, in her royal dignity and her contested legitimacy. The insult she never forgave. It poisoned the relations of the sister Queens when Marie, after the untimely death of her boy husband, returned to Scotland, and the long-drawn-out contest began, only to be terminated tragically at Fotheringay in 1587.

The Reformation had struck its roots deeply into Scotland, and the Protestants looked to Elizabeth for support. This she gave both in money and advice. Neither from a Scottish nor an English point of view can the proceedings of Elizabeth's Council on the one hand, or those of Murray, Morton, and their Protestant friends on the other, be defended, but from that of political expediency they can be grudgingly justified. They were in the manner of the times. The practice of statesmen to accept foreign bribes was one of the most detestable customs of the period. But it is clear that Elizabeth's Council could not afford to see a Catholic Scotland under French patronage, and that the Scottish Protestants could not be sure of their possessions, or even of their lives, if the old faith were restored in the North. It was a struggle for life, and the instinct of self-preservation drew all Protestants together.

When the question of a second marriage for the widowed young Queen of Scots was raised, and foreign suitors entered the matrimonial lists, Elizabeth and her Council were greatly concerned. They could not tamely permit the union of Scotland with France, Spain, or Austria through a marriage. They detested the idea of the union of Marie with any Catholic, whether foreigner, Scot, or Englishman, knowing the strong hold which the old faith still possessed over a majority of Elizabeth's subjects, particularly in the North of England, where a junction of English and Scottish Catholic adherents could be most readily effected.

The question why the ' long lad ' Darnley was permitted, in these circumstances, to go to Scotland with his father, the Earl of Lennox, when Elizabeth, according to Melville, had some suspicion that he might prove a fortunate wooer, is not easily answered. Possibly Darnley appeared so much less dangerous than a foreign prince that his appearance at Holyrood was rather a relief than otherwise, and Elizabeth may have believed in the assurance of the Lennoxes that they only wished to go North to look after their estates. Marie Stuart and Darnley had not then met. A marriage was only a vague possibility. It was not until the engagement was announced that its full import dawned upon the Council, and, had not the lovers hastened their marriage, steps would certainly have been taken to prevent it. Superior to Marie in experience and statecraft, Elizabeth was inferior in rapidity of decision. Cecil's natural timidity, Walsingham's prudence, and Elizabeth's love of long balancing of advantages, always militated against rapid action, and Marie Stuart was a woman who acted, sometimes unwisely, but with the rapidity of thought.

Why should Elizabeth have put forward her own favourite, Leicester, as a candidate for Marie's hand? The action was probably initiated by Cecil, who was hostile to Leicester's desire to wed Elizabeth, and may have thought that the Scottish marriage was an excellent scheme for killing two birds with one stone. But who shall venture to fathom the secret of Queen Elizabeth's impenetrable heart? She may have desired to see Leicester's son by Marie heir to the English throne. She may have shared, without acknowledgment, Cecil's point of view. Leicester's acceptance by Marie would have closed the book of Marie's suitors, to the very great relief of both Elizabeth and Cecil, and the Queen knew that Leicester would still remain subservient to her and never risk the confiscation of his great possessions. The only thing certain is that Leicester obeyed the Queen's orders with the utmost reluctance, and never pressed his suit. He was certainly wise. Among the Scottish nobles he would not have lasted long. All the Scottish leaders of those days died violent deaths.

Marie Stuart was never likely to have accepted Leicester. He was not of her faith. His attachment to Elizabeth was a matter of public notoriety, and there was much excuse for Marie's jibe that before Elizabeth offered her favourite she should marry him herself. Marie was already deep in her plans to combine the Catholic Powers in her favour, and the supple Rizzio had already become a useful instrument for the advancement of plans which Leicester would never have favoured. Without laying too much stress on Leicester's religious opinions, we must allow that he stood in the Protestant camp.

The Council met to examine the Scottish Queen's marriage plans on June 4, 1565, influenced, no doubt, by Randolph's reports of April and May from Edinburgh, wherein that competent ambassador kept Cecil fully informed of the first flirtations of Marie and Darnley. At this Council there were present the Lord Keeper, the Lord Treasurer, the Lord Admiral, the Lord Chamberlain, and the Vice-Chamberlain, three Earls, and three minor members, besides the Secretary, making thirteen persons in all, the normal Cabinet of Elizabeth. Foreign policy was not a hole-and-corner business in those days. Each problem was well thought out by the Council as a whole. The two questions put to them were, first, what perils might ensue to the Queen's Majesty, or the Realm, from a marriage between the Queen of Scots and Lord Darnley, and, secondly, what were meet to be done to avoid or remedy the same.

The Council advised that by such marriage a great number of persons in England might be alienated in their minds from their natural duties, and brought to believe that the succession of both

Crowns would be established in the Scottish marriage, so that they might favour all devices and practices tending to the advancement of the Queen of Scots. The Council held that, as the marriage was founded on trust in such as were Papists, it was also plain to be seen that these would favour the marriage and would use all means and practices to disturb the estate of the Queen's Majesty and the peace of the Realm, even by force, thus furthering the pretended title of the Queen of Scots not only to succeed the Queen's Majesty, but to supplant her. The Council said that intention and will are most manifest when power is greatest, and the contrary when power is less. So, they urged, when the Queen of Scots was Queen of France everything was done to impeach and dispossess the Queen's Majesty by writing and publishing that the Queen of Scots was Queen of England, by granting Charters and Patents, and by bearing the arms of England.

Darnley's party in England, continued the Council, was also expected to diminish the power of the Realm, which action the Council feared, because, they said, foreign princes had never prevailed in this Realm but with the help of some at home, a sentiment suggesting a deathless couplet of Shakespeare. The faction that favours the Scottish title, said the Council, is grown stout and bold, and might grow so great and dangerous that redress might become almost desperate. And to this purpose it must be remembered, they added, 'how of late in perusing of the substances of the justices of the peace in all the counties of the Realm, scantly a third was found fully assured to be trusted in the matter of religion, upon which only string the Queen of Scots' title do hang.' So the Council concluded that the perils were plain, and made so apparent by many sure arguments that no one in the Council could deny them to be many and dangerous.

So much being decided, the Council turned to what was to be done. They thought, in the first place, that Queen Elizabeth should marry and make therein no long delay ; that the reformed religion should be advanced, established, and fortified, both in

Scotland and England, and the contrary faith diminished, weakened, and enfeebled ; and that sundry things should be done to disappoint and break the intended marriage, or at least to make it less hurtful. The Council gave many particulars of the action to be taken in religious matters, and concluded by giving twenty suggestions for steps to break or to avoid the marriage. These were mainly of a military character, but also included the imprisonment of Lady Lennox, who was in England, the recall of her husband from Scotland, the seizure of his lands if he disobeyed, the strengthening of the links with the Scottish Protestants, and so forth. Finally, the Council said that, these advices being considered by Her Majesty, it may please her to choose which of them she liketh, to put them into execution by deeds, and 'not to pass them over by consultations and speeches.' No time, they urged, can so well serve as now, before the Queen of Scots' purposes are fully settled.

Elizabeth began to act upon this advice and sent Lady Lennox to the Tower, but before much more could be done Randolph reported in July that the marriage had already taken place, and that the happiness of the Queen of Scots with her husband was ideally perfect. So Elizabeth decided to make the best of things, relations temporarily improved, and eventually she became godmother or 'gossip' to Marie's son, the future James I. of England. However, the incident discloses the manner and methods of the Council's work, and the very thorough manner in which State affairs were not only discussed but promoted by immediate executive action. The gentle hint conveyed to the Queen that she could not afford to procrastinate showed that the Council knew her besetting sin in State affairs, of which they had had much evidence before, and were destined to have much more later.

An Anglo-French Treaty

If we turn to foreign affairs we can take as a typical example of Elizabethan practice the attempt made in the year 1581 to negotiate an Anglo-French Treaty, and to combine it with the marriage of Elizabeth to François, Duc d'Alençon, who became Duc d'Anjou when Henri, Duc d'Anjou, succeeded his brother Charles IX. as King of France. There is no doubt that a treaty with France was earnestly desired by the Queen and her Council at this period. The dangers from Rome and Spain were patent, and nothing but a French marriage and a French treaty promising mutual support appeared likely to Elizabeth's councillors to enable them to meet with confidence the dangers ahead.

A marriage with Henri, Duc d'Anjou, had been discussed since 1570, but when this prince proved recalcitrant his mother, Catherine de' Medici, substituted his brother François as a suitor. Elizabeth was then getting on in years. D'Alençon was nineteen years her junior. He was short, yellow, deeply marked by the small-pox, and as different as could be from Seymour, Courtney, Pickering, Leicester, Hatton, and other favourites who had basked in the sun of the Queen's favour, and were all tall, elegant, and handsome men. D'Alençon has been ridiculed by most British historians of this reign, most unfairly, for he behaved admirably throughout, was a most persistent suitor, and the Queen, carried off her feet, it would seem, by his gallantry, or by that of his agent, Simier, who opened the attack, fell in love with him for a time, or appeared by her conduct and her extraordinary letters to have done so. But in the end she tired of him, declared that she would not marry him to be Empress of the World, and threw on Cecil the ungrateful task of getting the prince out of her realm at any cost.

The Queen's English favourites had been in a terrible turmoil, but never really had cause for serious anxiety. A French marriage was bound to meet with insuperable religious objections. Leicester probably knew it, and Elizabeth must have known it too. If she carried it on so long we can only suppose that it was on account of the mischievous joy she took in love-making as an encounter of wits, and it may be that Anjou, and certainly Simier, behaved with more fervour than her calculating English favourites. No more gigantic royal comedy was surely ever played than during these years 1570-1581, when the wisest heads on both sides of the Channel were striving without cessation to bring about a marriage which, by the nature of things, could not take place.

But the treaty was the thing. Elizabeth was not deflected, even by the horrible massacre of the Protestants in France in 1572, from continuing to entertain a very real friendship for France. The associations of her mother with France may have had some influence upon her, and reasons of State perhaps more, but her real inclinations undoubtedly gave preference to a French rather than a Spanish or any other alliance. It was her considered view that 'whensoever the last day of the Kingdom of France cometh it will undoubtedly be the eve of the destruction of England.' She would never share the unfriendly feelings towards France which prevailed among her subjects, nor believe with them that France could never abstain from wars for three years together. She was incensed with the saying of Charles of Burgundy that 'the neighbouring nations would be in a happy case when France should be subject, not to one sceptre, but to twenty petty kings'; and when bad councillors advised her to seize Normandy and Picardy while Spain and the Leaguers were ravaging France she rejected the advice with indignation. 'Forsake not an old friend,' she once wrote to Henri IV., 'for a new one will not be like him. A bundle of rods bound together is not easily broken. There is no easier way to overthrow us both than by parting and disjoining us one from another.'

So when the Instructions to the English Commissions were drawn up on July 22, 1581, we see the pro-French sentiment of the Queen very plainly in the preamble:

> You shall move an assent that from henceforth we and the King [of France] shall, during our lives, be united in heart and minde, for confirmation of our honors, persons, states, dignities, kingdoms and dominions to either of us belonging; so that from henceforth we and the King shall be against all persons friends to friends, and enemies to enemies, notwith-

standing any former Leagues or Confederations with any other, or notwithstanding any counsel, persuasions, or motions to be made to either of us by any potentate or person, spiritual or temporal, to the contrary of the tenour of this League.

There were five articles in the English draft treaty. By the first, the contracting Powers agreed to support one another if attacked and to make the cause of the Ally their own. By the second, no traitors or rebels of either country were to be harboured in the territory of the other. The third article stipulated that no separate peace should be made with an enemy. The fourth laid down that if a larger force than 2,000 lances and 6,000 footmen should be required by either side, the party not attacked should yield all succour possible, but 'at the reasonable charges of the party invaded that shall so require a further aid.' Lastly, the fifth article declared that neither party should give aid to the enemy of any other prince with whom they were now at peace.

After a month of discussion, the negotiators began to discriminate between 'a treaty defensive' and 'a treaty defensive and offensive.' In a treaty defensive, mutual defence was arranged at the expense of the party requiring aid, in what manner, and in what time. So many soldiers, so many ships, victualled for so long. The aid provided was to be kept up so long as the party requiring it stood in need of it and paid for it. Powder and shot were allowed to be purchased in the allied country.

The offensive and defensive treaty laid down that upon invasion made by any third prince the confederates were mutually to proclaim him an enemy, and were then to invade his territory in such and such places, with such and such numbers, and at the expense of both parties, neither making peace except with the consent of the other. Free passage by the confederates through each other's territories was to be guaranteed.

Questions then arose whether definite numbers should be stated in the treaty, and how much should be paid by the party requiring them. The Queen was against the mention of any numbers as she did not wish to be bound by such an engagement, while Walsingham was for stating the numbers, because 'there hath risen always great controversies about these circumstances' in the past. There were various revisions in the drafting, and then the Queen proposed a new article :

> Item, to covenant with the French King, that he shall not give open aid to any enemy of the King of Spain, thereby to provoke him to make war upon the said French King, but that he shall first advertise her Majesty thereof, and have her allowance, otherwise her Majestie shall not be bound to aid the French King, or to make war upon the King of Spain. In like manner the Queen of England shall not, etc. *Ut supra, mutatis mutandis.*

This was practically Article 5 redrafted and made quite clear.

The treaty fell through, for reasons not worth discussing here, but the ideas at the back of the plan come out very clearly. It was to be for the lives of the two contracting monarchs and for mutual defence. The prince requiring aid was to pay for it, unless the enemy was invaded, when the cost was borne by the two confederate princes. The Queen would not have the number of the contingent fixed, and she took care that the French King should not drag her into war surreptitiously against Spain. Had this treaty been signed, and effectively carried out by both parties, England would not have confronted Spain alone as she did in the Armada year, while France would have been spared part of the fearful losses entailed by Spain and the Leaguers. In the thirty years of religious wars and invasions which ended with the Treaty of Vervins the population of France was reduced from some sixteen million to twelve million souls. The failure to complete the Treaty of 1581 must be ascribed solely to the French King, whose most inadequate reasons for refusing to sign will be found by the curious in the letter addressed to Cecil by Walsingham, Cobham, and Sommers, and sent from Paris on August 27, 1581.

Comments

What can we carry forward from these Elizabethan practices ? First, that complete education and ripe experience are necessary for the conduct of international relations. We can say that in these respects Elizabeth and Burghley were completely equipped and were superior to either France or Spain in the directing minds of those times, while we have only to read the English books and correspondence of the sixteenth century to be aware that the Queen and her Lord Treasurer did not stand alone in their comprehension of the arts of government. Under democratic systems such training and experience may be difficult to obtain, but at the same time no substitute can be found for them.

The Council in Elizabeth's day was practically in permanent session. The aid of Parliament was not invited nor desired in great affairs of State. The Council, and particularly the Secretary, were well informed by carefully chosen ambassadors whose couriers covered distances with great speed, while secret service was extensively used, and the reports of travellers supplemented official despatches. Special missions abroad were confided to great nobles and others who could speak the language of the country to which they were sent, and were schooled by travel in the manners and customs of foreign Powers. The records of all transactions were carefully kept. Cecil's method of laying before the Council a *questionnaire* (as we should call it now), and of inviting replies to each point, produced decisions as far as the Council was concerned, and it then remained for the Queen, if she had been absent from the Council, as she often was, to notify her assent or dissent.

Those who have leisure to examine a series of reports of these Councils will not all share the disapproval of the Queen's indecision and procrastination of which we find so many examples in contemporary letters. Elizabeth was very far-seeing. She had many advisers outside the Council, notably in the Church dignitaries, whilst the gossip of the world, carried by her ladies-in-waiting, penetrated to her Court. She had a large correspondence, and her penetration into the meaning of things was profound. Many cases submitted to her for urgent decision settled themselves by mere lapse of time, while as for her long dallying with foreign suitors, she seems to have understood better than her advisers that if matrimonial arrangements remained in the stage of negotiations the country of the prince making his suit was inhibited from doing England harm.

Would anyone now, looking back, wish the Queen to have married a foreign prince ? It is hard to believe it. A successor had come with the birth of the son of the Queen of Scots, and the best successor of all, since he was heir to the two Crowns, and effected that union for which the wisest heads in both countries had long sighed. Elizabeth had very good reasons for not actually appointing her successor, and facts in the end proved that she had been right in refusing, and that her Council and Estates had been wrong in so constantly urging her, to marry or name her heir.

In matters of general doctrine the Queen and her Council pursued the system of balance between France and Spain that had been established by Henry VIII. and Wolsey, and had long served, in another sphere, to preserve the security of Italy by holding the scales between Naples, Florence, and Milan. It was a delicate business for Elizabeth, because in those days England was numerically weak compared with the other two great Western Powers. But England's insularity, the renown of her old wars in France, and the military forces which the Queen could rapidly levy on land and sea gave her a privileged position. The danger of a Catholic Confederation was avoided, partly by the desire of France and Spain, to keep each other weak, and partly by the balancing ingenuity of English statesmanship and the unofficial war it waged in favour of the Protestant subjects of her two rivals.

It is one of Elizabeth's claims to honour that she kept the peace. It was not an easy thing to do in the sixteenth century. It was done because Elizabeth hated war, found the prosperity of England sufficient for her glory, and never hankered for annexations abroad. Cecil was pre-eminently a man of peace, and so was Walsingham. Most of the Council were of the same mind unless it was a question of supporting Protestantism, when they were all prepared to take considerable risks. There was no militarism in England in the Queen's long reign. There were many good leaders like Sussex, Hunsdon, and Mountjoy, and many fine seamen who made history, but no one presumed to interfere with the Council nor to deflect or influence the decision of the Queen in questions of war and peace. The people were ready enough for a fight if one was offered, and flew to arms as the Armada came up the Channel, but otherwise remained content with their new prosperity. It was England's sober statesmanship that enables us to look back with pride to the great Elizabethan period.

CHAPTER II

The Old Diplomacy: (2) Marie Stuart

Personal Appearance—The Enchantress—Activity, Courage, and Seduction—Coolness in Danger—Her French Upbringing—The Secret Deeds—Little Loyalty to the Crown in Scotland—Bothwell—The Queen's Arrival in Scotland—Efforts to be Named Elizabeth's Heir —Rivalry of the Two Queens—The Darnley Marriage Politically Considered—Murray's Revolt—He is Driven Out of Scotland—The Rizzio Murder—The Darnley Murder—The Protestants Prevail—Marie Stuart in England—The Trial at York and Westminster—The Queen of Scots Still a Danger to Elizabeth—The Crisis of 1570—Norfolk and the Ridolphi Conspiracy—Diplomacy and Courtship—Marie's Secret Correspondence—The Babington Plot—A Question of Marie's Life or Elizabeth's—Diamond Cut Diamond—Unanimity of England in the Armada Year a Political Justification for Marie Stuart's Execution—The Mighty Chessboard.

Her face, her form, have been so deeply impressed upon the imagination that, even at the distance of nearly three centuries, it is unnecessary to remind the most ignorant and uninformed reader of the striking traits which characterise that remarkable countenance, which seems at once to combine our ideas of the majestic, the pleasing, and the brilliant, leaving us to doubt whether they express most happily the queen, the beauty, or the accomplished woman. Who is there that, at the very mention of Mary Stuart's name, has not her countenance before him, familiar as that of the mistress of his youth, or the favourite daughter of his advanced age? Even those who feel themselves to believe all, or much, of what her enemies laid to her charge, cannot think without a sigh upon a countenance expressive of anything rather than the foul crimes with which she was charged when living, and which still continue to shade, if not to blacken, her memory. That brow, so truly open and regal; those eyebrows, so regularly graceful, which yet were saved from the charge of regular insipidity by the beautiful effect of the hazel eyes which they overarched, and which seem to utter a thousand histories; the nose, with all its Grecian precision of outline; the mouth, so well proportioned, so sweetly formed, as if designed to speak nothing but what was delightful to hear; the dimpled chin; the stately swan-like neck—form a countenance the like of which we know not to have existed in any other character moving in that high class of life where the actresses as well as the actors command general and undivided attention.—'*The Abbot*,' WALTER SCOTT.

No writer has improved upon Scott's description of the personal appearance of Marie* Stuart. He recalls the dead to life. We can recognise his accuracy as we gaze at the recumbent figure of the Queen of Scots in the Henry VII. Chapel of Westminster Abbey, or at Janet's famous miniature, or at the very few trustworthy portraits of the Queen that are commended to us by such accepted connoisseurs as Scharf and Lionel Cust. We are confirmed in our impression by the contemporary descriptions of French, Scottish, and English observers. We may, indeed, reckon Brantôme, Ronsard, Du Bellay, and Chastelard to have been courtiers as much as poets and writers, yet when we count up the victims who fell before Mary's charms during her meteoric career, we must certainly credit her with uncommon fascination and entrancing femininity.

The Enchantress

She was in truth an enchantress. She was the fine flower of elegance and distinction, and if this fact was undisputed in the glittering Court of Henri II., how much more must this feminine paragon have astonished the rough and warlike nobles of Scotland, and the burgesses as well, when the wheel of fortune brought their Queen back to Holyrood from France. She was also the best-dressed woman in Britain. Many of her letters testify to the attention which she devoted to her adornment, while her perfect taste and her skill in needlework supplemented the art of her Paris milliners, and her right royal jointure as Queen Dowager of France gave her the means to outshine all rivals at her Court.

Add to these advantages her inimitable grace, the dignity of her carriage, her winning familiarity of manners and address, her ready and persuasive eloquence, her pretty French accent, a voice that was, like so many French voices, peculiarly bewitching,

* The Queen herself always signed 'Marie,' and so did her four Maries. State documents conform, and the coins have 'Maria.'

and a remarkable capacity for weeping on the slightest provocation without spoiling her looks, and we cannot fail to see before us a strangely seductive feminine character. After all these years Marie Stuart's personality still exercises a despotism over opinion, makes many despise, denounce, or deny plain evidence, creates scores of posthumous lovers like Prince Labanoff, and inspires many loyal Scots to carry on, with the help of Jesuit fathers, an unending and furious battle with the Puritans and the disciples of Knox over every trifling incident of history that occurred 350 years ago.

But Marie Stuart was more than an attractive princess. She was superbly courageous, as courageous as Elizabeth herself; and were she living now she would certainly be one of the hardest riders to hounds in the kingdom. Those successful forays against Huntly and Murray which she led with pistols at her saddle-bow, and with all the vigour and unresting audacity of her brilliant and enterprising youth; that night ride from Holyrood to Dunbar two months before her son was born; the famous gallop from Jedburgh to the Hermitage and back when Bothwell had been sorely wounded by Elliot of the Park; and the sixty miles flight after Langside, show us that she possessed nerves and sinews of steel, and was never more at ease than in the saddle or happier than in the field. Her wit and coolness in danger were never more conspicuous than when she escaped from Holyrood two nights after the Rizzio murder, when the palace was garrisoned by enemies who included Murray, Morton, Ruthven, Lindsay, and the fiercest of the Douglases, nor was this the only occasion upon which she escaped almost alone from houses surrounded by her foes.

A daughter of James V. and a granddaughter of James IV. could scarcely expect to be exempt from love entanglements, but a descendant of the hardest of the Tudors was also bound to have a heart like a diamond when she hated. It is questionable whether she ever knew what fear meant till the beldames of

Edinburgh threatened her with the stake after Carberry Hill. She was extraordinarily rash, not to say reckless, in her intrigues when a prisoner in England, and the wonder is not that she was eventually given away by the clean sweep of her correspondence and ciphers at Chartley, but rather that eighteen years should have elapsed without any previous investigation of her papers. Let us spare our praise of Walsingham's trap which caught Marie Stuart at last. By her rashness and voluminous cipher writings the Queen of Scots was always asking to be entrapped. A less courageous soul would have been a shrewder plotter.

It is rather absurd to pretend that because the Court of Henri II. was, or was accounted, dissolute, Marie Stuart became contaminated by her French environment. The French kings knew too much not to know how to look after their young princesses. Marie was brought up with the royal children under the care of the learned and estimable Margaret, sister of the French King. There is not a shadow of a shade against Marie's character until, long afterwards, Darnley, by turning murderer and traitor too, awakened, to his cost, the previously dormant Tudor characteristics of the Queen.

Marie was well educated. In elocution she became remarkable. She had a particularly sweet voice, and accompanied herself on the guitar-like lute. Her versifying was respectable for a princess. She knew Latin well enough, and Italian too, while for intelligence and discretion she was awarded the palm among contemporary women by that astute statesman and excellent judge of merit, the Cardinal of Lorraine, who must be regarded as her earliest mentor in statecraft. The seven volumes of her letters which Prince Labanoff's industry collected are among the most famous written by any woman at any time. If she was taught neither English nor Scots in France, there was a political reason for the omission. The King of France wanted Scotland and Scotland's Queen to become French. Her first letter in English was written to Sir Francis Knollys from Bolton in 1568, after her flight to England, and truth compels me to admit that I am a little dubious how and where a Sheffield examiner would place her now.

> Mester Knoleis [she writes], y heuu har sum neus from Scotland; y send zou the double of them y vreit to the Quin my gud sister, and pres zou to du the lyk, conforme to that y spak zesternicht unto zou, and sut hesti ansur y refer all to zour discretion, and wil lipne beter in zour gud delin for mi, nor y kan persuad zou, nemli in this langasg; excus my iuel vreitin, for y neuuer used it afor, and am hestet . . .

Marie Stuart was more sinned against than sinning in the matter of statecraft in France. Henri II. gave the measure of French kingcraft when he and his counsellors induced Marie, before her marriage with the Dauphin, to sign three secret deeds, one of which made over Scotland in free gift to the Crown of France in the event of her death without issue. These deeds completely nullified the public contract which the eight delegates of the Scottish Parliament brought to France and signed in common with the young Queen, the Dauphin, and the King of France. For an instance of perfidy the act has few equals in history, and fully corroborates the view which Maitland of Lethington had formed when he reluctantly separated himself from the cause of Marie's mother, the Queen Regent, on the valid plea that France was exploiting Scotland in purely French interests.

Marie Stuart was little over fifteen years of age when the secret deeds were signed, and most Scottish historians have therefore absolved her from blame and responsibility for them. She herself in after-life always denied this responsibility, and it is a fact that she always displayed a strange submission to those in authority over here, whether guardians, lovers, or husbands, until they forfeited her good opinion. If Marie's subjects had just cause to complain of the treachery of 1558, these subjects played their Queen an equally shabby trick when the Scottish Estates sought to sell the crown to Elizabeth at the price of an Arran marriage. Only Elizabeth's refusal to wed brought the scheme to the ground.

The principle of loyalty to the Crown was not too remarkable in Elizabethan England. In Scotland it only existed when private interests coincided with royal policy. The steady loyalty to Marie Stuart of men like Seton, Herries, Leslie, and a few more forms the exception to prove the rule. James Hepburn, Earl of Bothwell, was also loyal to the Queen, as he had been to her mother, and the house of Hailes was hereditary friend to widowed Queens. Bothwell was an enemy of England, a man of reckless audacity, cultivated, and with brains. No one but a Danish biographer has given us a candid story of his life and conduct, and it is thought proper by most people to abuse him, and to travesty his appearance, his conduct, and even his age. He took part in the murder of Darnley, but when we reckon with the fact that half the Queen's Council had recommended that the 'young fool' should be got rid of, Bothwell's responsibility was no greater than that of the other lords who shared in the Craigmillar and Whittinghame conferences, but avoided the risk of attendance at Kirk-o'-Field. Bothwell was at least a man who never sought by subterfuges to escape the consequences of his acts. The idea that Marie Stuart could fall head over ears in love with the caricature of a man which most historians present to us under the name of Bothwell must be dismissed. Their Bothwell is not the Bothwell of life, but of prejudice.

Marie in Scotland

When Marie Stuart returned to Scotland after the death of her schoolboy husband, François II., she had determined to accept the Reformation on the understanding that she should be free to exercise her own religion. She preferred to be guided by Murray and Lethington rather than to throw in her lot with the Catholics, because it was her most ardent wish to be recognised as Elizabeth's heir. There was a brief period in 1562 when the success of this plan seemed not impossible, but, in spite of all the efforts of her

Protestant counsellors, the scheme fell through. The two Queens never met. The treatment of Marie by Elizabeth was harsh and unfriendly, but from an English point of view it was comprehensible. England alone had escaped from the religious wars which were desolating Europe in the sixteenth century, but more than half of the better elements in England were still Catholic. The appearance of the seductive Scottish Queen in London was not a risk that Elizabeth and her Privy Council were prepared to face.

Sir James Melville's description in his memoirs of his audiences with Queen Elizabeth at Westminster in 1564 is one of the most interesting glimpses that we possess of the English Court at the time. Elizabeth's interest in the appearance and charms of Marie Stuart was extraordinarily keen, and Marie's ambassador was much put about to answer the questions concerning the relative fascinations of the hair, complexion, height, dancing, playing, singing, and dress of the sister-Queens. To please Elizabeth without crying down the credit of the Queen of Scots needed diplomacy of a high order, but the obvious desire of Elizabeth was to know, if she met Marie Stuart, which queen would bear the palm for beauty and accomplishments.

The negotiations with Elizabeth, while they lasted, kept Scotland quiet, and the first years of Marie's presence in her realm were happy. It was only when the hope of being declared heir-presumptive faded away, and when the project of a marriage with Don Carlos had been wrecked by the Guises, that Marie turned to an English marriage. Her choice of Lord Darnley was politically well conceived so far as the plan went, for Darnley's father, Lennox, disputed with the Hamiltons the succession to the Scottish throne, while Darnley himself was next lineal heir after Marie to the English crown. But the fragment of a memorandum by Marie Stuart in the State Paper Office recounting the political advantages of a Darnley marriage does not at all refute the theory of a love match, even if Marie's caprice for the 'long lad' was fleeting. That Marie had, for a few months at least, a violent partiality for the youth whom the Cardinal of Lorraine described as a *gentil huteaudeau*, and Elizabeth as a *jeune perche à belle taille*, is proved to demonstration by the reports from Edinburgh by Randolph at the time. Darnley was a tall youth with a good figure. The first coins on which his head appears give him a handsome profile, and tell a story different from that of the best-known portraits of him. The defects of his character only appeared later. He was always too much of a cub for a woman like Marie Stuart.

Marie's Catholic nobles approved of the marriage, which was tolerated by several Protestant lords on account of their relationship with the Lennoxes. Some of these were subsequently among the Queen's bitterest enemies. But Elizabeth's Council was hostile to the plan, and it was in expectation of English support that Murray and several other Protestant nobles raised the standard of rebellion. Marie, for reply, liberated George, fifth Earl of Huntly, and 'shook Bothwell out of her pocket' at the rebels. Murray and his friends, all clients of Elizabeth, were driven out of Scotland, and in her triumph Marie determined to restore the Catholic faith, in which project she was strengthened by the promises of Philip and of Rome. That Marie joined the Catholic League at this time, as Robertson, Fraser-Tytler, Mignet, Froude, and other historians have asserted, I can find no satisfactory evidence to prove, and I take Dr. Hay Fleming's doubts on the accuracy of their opinions to be justified. But that from this time forward Marie Stuart was leaning on the support of the Catholic princes and receiving their help is not open to dispute. The war against Elizabeth, fought less with arms and armour than with brains and breviaries, had begun.

But let us admit that from Marie's arrival in Scotland in 1561 up to the close of 1565 she had been wonderfully successful and was greatly beloved. She could not win over the implacable and uncompromising Knox, despite skill in argument to which Knox himself has paid adequate tribute, but she had suppressed all her traitors and had made Scotland united and as happy as circumstances permitted. As she stood near the Border with nigh 20,000 men at her back, after the fierce excitement of the 'chase-about raid' after Murray and his adherents, she might well think and speak of leading her army some day to London. Fair seemed the prospect, for Scotland was superficially united, the English Catholics were expected to rally to her, while Rome and Spain were at her back, and France was friendly. These fond and delusive hopes were shattered by the sudden opening of a drama of love, hate, passion, and intrigue such as even Scotland had never known before. It began with the murder of Rizzio on March 9, 1566, and terminated for the time with the flight of Marie to England after Langside in May, 1568.

That drama does not concern statecraft, except that the murder of Rizzio was certainly caused by the apprehensions of the leading Protestants that their lives and fortunes, and those of Murray and the exiles, were not safe with the Catholic revival which they foresaw.

Once Marie Stuart determined to effect that revival, accepted that learned dialectician John Leslie, Bishop of Ross, as her ecclesiastical adviser, and dropped Lethington for Rizzio who was devoted to the cause of Rome, she raised up enemies who were fatally destined to overthrow her at last, so thoroughly had the Reformation taken hold of the mass of the Scottish people. Her best counsellors, including Lethington, would not follow her, and in Elizabeth the Scottish Protestant lords found a support which—uncertain, fickle, and tricking as it often was—in the end prevailed.

Elizabeth from 1560 onward had assumed a kind of suzerainty over Scotland, and found it only too easy, by a comparatively small expenditure of money, to harness many powerful but poor and avaricious Scottish nobles to the cause of Protestantism, which was also theirs. Marie Stuart had the misfortune of never, or never for long, being able to surround herself with a trusty body of counsellors to prosecute a settled policy. The Bothwell marriage, joined to the belief of the common people that both Bothwell and the Queen were responsible for Darnley's murder, ably exploited in both cases by the Presbyterians, turned the hearts of the townspeople from Marie Stuart. There is a gorgeous, reckless, uncalculating humanism about Marie that wins many hearts. But she was altogether too human, and in those harsh days of the Lords of the Congregation and the pulpit of St. Giles there was no pact with human weakness. Scotland had to be ruled, and ruled on Protestant principles. So Marie passed out of Scotland for ever, and though the 'Castilians' held Edinburgh till May 1573 under gallant Kirkaldy, with Lethington, who was always, openly or secretly, the Queen's friend, the royal strongholds were gradually reduced and the theatre of Marie's activities changed.

The Captive Queen

The cause of Marie Stuart as a Queen dethroned by rebels was the cause of every prince in Europe, not even excepting Elizabeth. If neither France nor Spain intervened during Marie's incarceration at Lochleven, it was because both were engaged at the time in their own civil wars, which absorbed all their attention and resources. Elizabeth, however, intervened with all her customary imperiousness, and to this intervention Marie owed her life. When again the inquiries took place at York and Westminster into the responsibility of Marie for Darnley's death, and even after Murray had been forced to show the originals of the Casket letters to the Commissioners, Elizabeth finally closed the proceedings by declaring that 'nothing had been sufficiently proven or shown by them (the Scottish Commissioners) against the Queen their sovereign whereby the Queen of England should conceive any evil opinion of her good sister.'

It is difficult to believe that this was her real opinion, for much evidence to the contrary is extant, but she had forced Murray to produce his proofs and had discredited the Queen of Scots. For Elizabeth that was enough, and she was not going to give the Scottish rebels the satisfaction of winning their case against a sister-Queen and of inflicting a wound upon the prerogatives of majesty. In considering the whole attitude of Elizabeth towards Marie Stuart from 1568 onward we must remain conscious of the ceaseless bombardment of London by the European princes in favour less, at first, of Marie herself than of the monarchical principle, which had been outraged by her treatment by her subjects. To such appeals Elizabeth could not remain deaf or insensible, and they account for the long irresolution of the English Queen even after the verdict of Fotheringay.

It does not concern my subject to express an opinion on the authenticity or otherwise of the Casket letters. The originals disappeared after the execution of Gowrie, from whose hands Robert Bowes sought unsuccessfully to obtain them for Elizabeth ; so the defenders and the enemies of Marie Stuart can wrangle to their heart's content over the letters to the end of time without convincing their adversaries of their error.

It is a mistake to suppose that Marie, the captive, was no longer a danger to Queen Elizabeth. She was, on the contrary, a greater danger when captive than when free. From the moment of her imprisonment in England she became the Catholic martyr, and her cause constantly progressed amongst the Catholics of Europe. The whole strength of her remarkable ability aimed at the combination of the Catholic princes in Europe with her adherents in Scotland, and with the Catholic subjects of the English Queen. We are constantly and disagreeably surprised to find the names of so many of the English nobility among the participants in Marie's plots. Her projected marriage with the Duke of Norfolk was planned to conciliate these nobles, but while Rome was willing to annul the marriage with Bothwell on the plea of the rape of

Almond Bridge, the Scottish Convention at Perth broke up without sanctioning annulment, and as the marriage had taken place under Protestant rites, Rome could not dissolve it.

Marie, the mother of Great Britain, died Bothwell's widow. The Bothwell marriage was too good a card in the hands of Elizabeth and Murray to permit either to part with it. Still the project of the Norfolk marriage held the field, and while Spanish money was lavishly expended among English Catholics by Don Gueraldo d'Espès, Philip's ambassador in London, Pope Pius V. spared no pains to help the cause, and de la Mothe Fénélon, French ambassador in England from 1568 to 1575, constantly urged Elizabeth, in the name of the French King, to restore Marie to her throne.

The Crisis of 1570

It was a perilous position for Elizabeth. She had aided the rebels of France and Spain, and was not in the good graces of either Power. The danger was greater because she had so long been kept in ignorance of the defection of some of her own subjects. It was not only that Norfolk himself was the highest, greatest, and most powerful of her nobles : there were too many other English nobles in the South who supported him or tacitly assented to his scheme, while the Percys, Nevilles, Nortons, Dacres, Tempests, and others in the north were ready to rise when Norfolk gave the signal. There followed, when Leicester gave the secret away, an outburst of Tudor rage which must have made Elizabeth's counsellors all tremble for their heads. Had Norfolk possessed a fraction of Bothwell's enterprise or intrepidity it might have gone hard with Elizabeth, but he was timid in action, and surrendered at the Queen's command. Norfolk was a broken reed. The Pope was too slow. Philip was too distant, and Alva unready. Elizabeth was about to begin those fine-spun and lengthy flirtations with the French princes which kept France off her back almost to the eve of the Armada.

Let us, however, recognise how great the danger was. Had the Pope's Bull of Excommunication been antedated by a twelvemonth, the year 1570 might easily have proved fatal to Elizabeth. Even as things were, the expected restoration of Marie Stuart, combined with the murder of the Regent Murray by Bothwellhaugh, had drawn her old Scottish adherents into the field again, and it required the savage invasion of Sussex and Scrope to suppress them and to re-establish the young King's party in Scotland, while the English rising in the North bore at first a formidable aspect. Had Alva threatened a descent, the march of the English troops northward would not have been practicable.

Things turned upon the throw of a die. Marie Stuart, a prisoner in England, had found that the call of the old faith was the most potent of all charms to unite the Catholics of the world in her cause. The attempt of 1570 failed rather from want of direction and combination than from want of means. Not at Langside, but in the great failure of 1570, the cause of Catholicism was lost in England.

As we meditate over the whole position we may fairly question whether the imprisonment of Marie Stuart in England was sound policy at any time. We have the feeling that had Elizabeth died from any cause the bulk of the nobility and gentry in England would have flocked to Sheffield Park to acclaim Marie Queen, just as they had flocked to Hatfield to acclaim Elizabeth before Mary Tudor's corpse was cold. Marie Stuart was not only the hope of Catholicism in Britain : she was the magnet which attracted all the restless and discontented spirits in the kingdom. In France she could have done little harm, for Catherine de' Médici's life was spent in destroying Marie's hopes, while Spain would never have stood a Marian invasion of England with French help. Marie Stuart in England was always a call to arms for all Catholics, and a very serious danger to Elizabeth's life. She became the stormy petrel of British politics, and the harbinger of tempests in the State.

The Last Phase

After the Ridolfi conspiracy and the execution of the Duke of Norfolk, Marie turned from France to Spain, and Philip became her most ardent supporter. He must have remembered 1554, and the bribes by which Egmont had so easily won over waverers among the English nobility to the Spanish marriage. Marie might have preferred France, but that superb stroke of statecraft which led Elizabeth to open the negotiations for a French marriage, and the inimitable skill with which she prolonged the flirtations for nearly twelve years at a moment when a Franco-Spanish combination would have been perfectly fatal to her, placed France out of court as Marie's ally.

The danger for England now somewhat altered in aspect. The usual Protestant affirmations that the Pope's excommunication of Elizabeth in 1570 had no effect are not tenable. It authorised Catholics to transfer their allegiance from Elizabeth to Rome, and the consequences were very quickly seen. From that date we trace all the most dangerous plots against the lives of Elizabeth and her counsellors, plots made and engineered mainly by English seminarists trained abroad, who found too many and too willing tools among the gentry in England. These plots were defeated by the sleepless vigilance of the gloomy and pertinacious Walsingham and his secret agents, but we cannot admit that murder plots would have attracted such men as Babington and his more or less aristocratic comrades had not one and all held themselves to be justified by the Bull in transferring their allegiance to Rome.

Throughout all the last period of her life Marie Stuart had scarcely anyone with her who was fitted to be her counsellor. But she had from a dozen to thirty trusty attendants, secretaries, and servants who organised a secret correspondence of a very efficient kind. Letters were exchanged with the Courts of France and Spain, and with Rome, while the Catholic ambassadors in

London, the Bishop of Ross, and Marie's other emissaries, kept their mistress regularly informed, and touch was also preserved with Scotland and with the many adherents of Marie in England. The Queen of Scots was becoming desperate, and caught at every chance, no matter how hazardous, which seemed likely to lead to her liberation. The central features of the Babington plot were the murder of Elizabeth and a Spanish invasion. In an evil moment Marie Stuart received a letter from Babington, and answered it through her secretaries ; so at least it was alleged at her trial, when the deciphered letters were produced, and formed damning evidence against her, but she herself denied that she had ever written to him.

The conviction firmly implanted in the mind of every Protestant and Catholic in England and Scotland that it was a question of Marie's life or Elizabeth's, and that neither could be safe while the other lived, accounts for the bloodthirsty sentiments of 1586 which were common to Catholics on the one side and to the Protestant, verging on Puritan, sympathies of the English Parliament and Privy Council on the other. The Protestants made reiterated efforts to coerce Elizabeth into drastic action against the Queen of Scots, while the Catholics were prepared to do away with Elizabeth and to greet a Spanish invader with joy. It was diamond cut diamond, and the horrible letter sent by Walsingham and Davison to Paulet, suggesting that he should murder Marie, as well as Leicester's characteristic suggestion from the Netherlands that she should be poisoned, show to what length religious passions and political fears had brought normally sane people. It was a question of life or death for many beside the two Queens, and such words as mercy and moderation had been expunged from the vocabularies of the disputants.

The Armada came too late for the Catholic cause. Things which English Catholics would have been ready to do for Marie Stuart they would not do for a foreign prince. The unanimity of Britain in 1588 was the political justification for the terrible scene enacted when that disrobed and stately figure of the Queen of Scots, stripped to her blood-red bodice and skirt of velvet and taffeta, and standing silhouetted out against the black background of the block and the scaffold, made the last atonement, and died as she had lived, with royal dignity and the calmness of a great courage.

It was a mighty chessboard on which the fate of Marie Stuart was played out. There were the Blacks and the Whites, with the Kings and Queens, the Knights, the Bishops, and the Castles not to speak of the Pawns—the poor little Pawns of the common people, of whom little account was taken in the sixteenth century, though they fought valiantly enough and were sacrificed remorselessly in the interest of the greater pieces. But the great pieces fought too. They paid in their persons. Nearly all the chief pieces on the Scottish side of the board were swept relentlessly off the table by deeds of violence, and perished during the terrible struggle which established Protestantism in Britain at the price of blood and sacrifice.

CHAPTER III

Disarmament and the State of Europe*

A Summary for American Readers of the State of Europe—The Washington Conference—The Question of Disarmament—The Victors still Dominant and the Vanquished still in a State of Subjection—No Satisfaction to the Victors to Remain Armed—A Political Necessity—Causes for Disquiet—Difficulties Surrounding Disarmament—A Statement of the French Case—The Loss of the Anglo-American Guarantee—The Case of the Other States—Alliances and Understandings—The First Hague Conference—Obligatory Arbitration—The League of Nations—The Sorrows of Europe—The Question of Exchanges—Currency Reform—Questions More Vital to England and America than even to Europe—Obstacles to Trade—The Succession States of Austria—All Governments Want Money—The New Agrarian Policy in Europe—Reaction Against Bolshevism—German Politics—The Real Diseases of the World.

All Commonwealths ought to desire Peace, yet it is necessary ever to be prepared for the War ; because Peace disarmed is weak, and without Reputation : Therefore the Poets feign, that Pallas the Goddess of Wisdom did always appear armed.—Sir Walter Raleigh, "*The Arts of Empire.*"

The Washington Conference is about to open, with disarmament for its leading theme, and I think it may be interesting to American readers if I give them, for what they are worth, the deductions that I have drawn concerning disarmament and kindred subjects during recent travels from the Baltic to the Ægean and from the Channel to the Black Sea. These journeys have occupied me during the greater part of this year, and have brought me in contact with most of the directing minds which exercise authority

* This article appeared in the *Atlantic Monthly* for Nov., 1921. As it represents my opinion of the state of Europe to-day I have made no changes in it.

in the old Continent, as well as with many other people of all classes, professions, and nationalities. I write for American readers with the greater pleasure because, wherever I have been, I have found English and American opinion firmly united, with or without previous discussion or agreement, on almost every single question that distracts Europe, and I have certainly returned home with this fact as the most satisfying, if not the only satisfying, conclusion of my tour.

The Question of Disarmament

One may divide Europe, broadly speaking, into three parts: the victors, the vanquished, and the neutrals in the late war. The victors are suffering from indigestion, the vanquished from exhaustion, and the neutrals from the discomforts inherent in propinquity to sick neighbours. No people are happy; no nation loves another; and it will take years for the hates and jealousies arising out of both the war and the peace to die down. Practically speaking the victors are still dominant and the vanquished still in subjection. The victors are dominant because they are compelled, in greater or less degree, to remain armed until all the terms of the peace treaties are carried out; and this must be an affair of long years, because the reparations exacted, though not a tithe of the real cost of the damage done, have been spread over long periods of time, in order to make the payments possible. The presence of numerous Inter-Allied commissions in the conquered countries is a source of humiliation to them, but cannot be helped, as they are there in pursuance of treaties.

It is no satisfaction to the victors to remain armed, because the cost is great and every state is at its wits' end for money. In fact, the destitution of treasuries is so marked that even the victors have to impose on their own people almost unendurable burdens, and in many cases do so with little regard for the elementary principles of economics, thus helping to prolong the crisis of which even America is sensible. But they dread that, if they do not remain armed and impose these burdens on their taxpayers, the vanquished may either recover and renew the war, or, at all events, find good pretexts for discontinuing their payments, owing to their recognition of the fact that there is no power sufficient to coerce them. In this event, certain of the victors will reckon themselves ruined.

Therefore, the first unpleasant fact to be faced is that the victors are still armed and the vanquished almost entirely disarmed ; and that, though this is an intolerable state of affairs, offers no permanence, and heals no wounds, an alternative is not within sight for many years without risk of the renewal of the war, which alternative is, of all things, the one that nobody can contemplate with equanimity. 'Peace disarmed' would be not only 'without reputation,' but a signal danger.

A conference aiming at disarmament will observe that, England apart, and America having side-tracked herself in this business, the victors retain compulsory service, while the vanquished, or at least their governments, all pine for such service and are not allowed to have it. Similarly, the vast war-material of the victors remains in existence, rotting or rusting in part perhaps, and gradually growing out of date, but still more or less fit for use ; while the huge war-material of the vanquished, greater by far than anyone imagined at the Armistice of 1918, has been swept into the net of the victors and has either been taken or destroyed. Disarmament ? Yes, it has been carried out by force, but only in the case of the conquered states.

Another cause for disquiet is the fact that practically the whole of the able-bodied population of Europe were trained soldiers in 1918, or trained organisers or providers of the needs of war, in one form or another. Therefore, if some strong compelling sentiment should make a people rise, it would only need arms for numerically strong forces to reappear as by magic, and all the long training of the war period could be dispensed with. This situation will not end for another fifteen or twenty years, when all the veterans of the war-time will be too old, or too stout, or two much immersed in their new occupations, whatever these may be, to desire, or to be able, to march and fight. The victors have seen very clearly that these veterans cannot be destroyed, but that war-material can be ; and the various Inter-Allied military commissions have therefore concentrated upon material, and have shown relentless severity in insisting upon a thorough surrender of arms—not only of guns and rifles, aeroplanes and machine-guns, but of the whole machinery of military equipment, including carts and limbers, harness, and all the thousands of articles that go to make up a properly found army. It is held that this action will make the vanquished states incapable of creating modern armies, except after a long delay, which the victors will naturally exploit.

The vanquished, on their side, have naturally sought, by every available means, to escape the control of the military commissions, and, in effectives as in armament, to conceal what they are doing by more or less clever camouflage. It has not succeeded, on the whole, but there are still military organisations in excess of treaty stipulations ; there are all sorts of pseudo-civilian societies, associations of old soldiers, compulsory-labour laws, and so forth, which are not indeed very formidable, but which show that the disposition endures to resuscitate military power at the first opportunity. Similarly, there is a certain amount of war-material still concealed and undelivered, especially rifles and machine-guns ; but to me the wonder is that so much has been given up, and I feel confident that it would not have been had the vanquished been certain allied and associated powers that one could name.

However, there it is, and that is the present situation. But not quite all has been said ; for it is the decided and well-weighed opinion of the best men in control of the military commissions that, after they withdraw from the territories of the vanquished states, it will not take more than two years for the war-material to be replaced, at all events in the case of Germany; and that in five years the whole of the vast war-material may be renewed, quite apart from contracts that may be made with neutrals, perhaps through foreigners. Therefore, the question arises whether these commissions should not be retained until all the veterans are past the fighting age; for though, by the Treaty of Versailles, it is the League of Nations that has the duty of checking future designs of an aggressive sort, the League will have difficulty in carrying out this task; and, in fact, no one believes that it can do it.

Another real difficulty is that, when we disarm a state, we practically become, in a moral sense, trustees for her internal order and external security. A country whose forces are compulsorily reduced to the vanishing-point may not be able to suppress Spartacists, Bolshevists, or what not; may not be able to prevent bandits from crossing from their territory into another, or to keep out other peoples' bandits; while there is the still more serious danger that the government itself may become so weak that it may lack authority, and be at the mercy of a *coup d'etat.* This lack of authority is one of the most constant complaints of the vanquished states.

It is certain also that a long-service, voluntarily enlisted army, gendarmerie, or police, offers an easier prey to intriguers than a conscripted army based on short service; for the latter constantly refreshes itself from the whole people, whence it springs, while a volunteer force has to be taken from less choice elements, and in unsettled times and territories easily becomes a sort of Prætorian Guard, or corps of Janissaries, at the call of the highest bidder. In countries of peasant proprietors it is even difficult to recruit a voluntary army at all.

These are among the problems that Washington will have to confront on the side of the recently vanquished states; but perhaps they will be surpassed in complexity when the armies of the Allies are passed in review.

It is true that England will not have much difficulty in securing a clean bill of health, because we have scrapped compulsion and all our military acts of the war period. Except for the possession of better material and equipment, and for the acquired precedent of creating a national army based, at need, on compulsion, we are in a worse state of military destitution than we were in 1914 — which is saying a good deal—whereas we have much greater commitments all over the world, and a whole series of new difficulties for which, in ultimate analysis, force may be the only remedy.

But when I think of our allies, they will, I imagine, be asked to explain their position ; and they may possibly be asked why, if the disarmament of their late enemies has been in such large measure accomplished, they do not themselves disarm. The retention, practically all over Europe except in the vanquished states, of compulsory military service, and of the potentially huge armies which derive from it, will not, I imagine, escape comment. The case of our allies I will, therefore, briefly state.

If we take France first, we must admit that she has the greatest, and, perhaps—with a saving clause for Japan—the only really great army in the world. She has a numerous, well-organised, and splendidly equipped army, much superior to her army of 1914, led by commanders of the greatest distinction, and capable, as I verily believe, of conquering Continental Europe. If a Bonaparte came into view, he would have a perfect instrument ready to his hand, with this reservation, that—at first, at all events—Frenchmen would not march except in a good cause, and with the object and scope of an operation clearly pointed out to them. But such eventualities are, I hope, far from us. French generals do not dabble in politics, and the whole army despises them. No political generals in France survived the war-storm. No civilian could, or would wish to, repeat the Napoleonic *épopée*, of which he would probably be the first victim.

But even more important is the fact that France's population is small, and that her strength to-day, admittedly great though it be, is merely a fortuitous and perhaps temporary superiority of an army, and not one of a people firmly based on foundations of superior numbers, wealth, and science. France might march on Berlin, even on Moscow, and reach both with ease; but she is quite incapable of confronting the subsequent hostility of the world, or even of Europe, which every aggressor must expect who attempts to emulate the projects of Napoleon or Wilhelm II. We must keep our heads cool when we observe the brilliant power of France.

The maintenance of the French army at its present standard of numbers and efficiency is due to want of confidence in the future; and if France pleads this want of confidence, one must be just to her and lay the blame where it is mainly due, namely, upon the lapse of the Anglo-American guarantee. France reluctantly consented to abandon her defensive plans on the Rhine because America, and England if America ratified the agreement, were to give France a guarantee against German aggression in the future.

Two years have passed, and America has not ratified that undertaking. Consequently our adhesion falls to the ground, although our Parliament accepted the liability under the conditions named. Very likely we on this side of the water were very great fools, and curiously ill-informed of the real state of public opinion in America, when we signed that conditional guarantee. That remark applies to our Government, if the cap fits them. It depends upon whether our former Ambassador at Washington warned the Government that the American Senate might not second the guarantee of President Wilson. I do not know whether our Ambassador gave a warning or not. But the public in England and France certainly never had the glimmer of a suspicion that a guarantee signed by a President of the United States, and countersigned by a Secretary of State, in a vital matter affecting the safety of France and the future peace of Europe, would not be honoured in America.

It is impossible not to attribute a very large share of France's want of confidence in the future to the above cause, and a very large share of Europe's unrest to France's want of confidence. Over and over again I have been told by French statesmen and generals that France would never have taken the unrelenting course that she has taken towards Germany had the Anglo-American guarantee stood. Over and over again I have been assured by representatives of all the allied and associated powers that Germany would never have dared to confront that combination, and that, secured by the guarantee, France would, and could safely, have disarmed. The fact that none of these things happened is the main cause of the sanctions, the Upper Silesian trouble, the reparation wrangles and most of the resulting unrest that has followed throughout Europe, which seems to take its cue from the barometer of Franco-German relations.

I am not blaming America in the least. Our own long-established practice to keep out of continental entanglements when we can, is as deeply rooted in principle as that of the United States to steer clear of Europeon commitments. The difference between us is merely the difference betweem the breadth of the Channel and the breadth of the Atlantic. By that much our policy differs from yours ; but it is a difference of degree, and not of kind. But for all that, when one observes, as every traveller through Europe must observe daily, the truly appalling results that have followed from this failure, misconception, desertion, or whatever one should term it, one stands aghast at the consequences, and laments the little wisdom with which the world is governed.

France has no definite guarantee now that any state but Belgium, and perhaps Poland, will support her when Germany feels strong enough to act ; and, in the sheer desperation of self-defence, has thought it necessary to inflict upon her neighbour one humiliation after another, in order to make her, and keep her, weak. The policy of broad and genial tolerance, which would have so well become a country with France's generous traditions, she could not follow, for with her forty millions there were over against her seventy million Germans, with a far higher natality; and France saw no salvation except in the rigid exaction of all her treaty rights, so that Germany, for a great number of years hence, might be inhibited from even dreaming of revenge. But when one thinks of the dry-powder régime under which France has been living for so long, and of all the terrible injuries inflicted on her by Germany in the past, one can understand, and *tout comprendre c'est tout pardonner.*

If France declares at Washington that nothing tangible except her army stands between the world and the renewal of the war by Germany, I do not know how she can be gainsaid. In the circumstances, it is the truth. I even think that we English and Americans, having left to France the largest share of the war, must feel a tinge of shame at leaving also to her the main burden of enforcing the peace, with all the obloquy that follows.

Italy will plead that she has greatly reduced her army and diminished the service periods. She can say with justice that her policy has been conciliatory, and that she has shunned adventures. But she can also show that the *Anschluss* movement in Austria has underlined the danger of Austria joining Germany, and she can point out that such an act would bring Germany down to her borders. Yugoslavia can urge that both Hungary and Bulgaria are uneasy neighbours; Czechoslovakia, that she is liable to be stifled by the Germans round her, and has Austria and Hungary to fear. Rumania can point to dangers from three neighbours, and, above all, from the Soviet armies upon the Dniester, and from the bulk of the Bolshevist reserves not far away.

Belgium has too complete a case to bring up from 1914 for anyone to find fault with her for abandoning her neutrality and reorganising her army on more modern lines; while Poland can say that she has recently saved Europe from the Reds by her military exertions. Lastly, there is Greece, who can show that she went to Asia Minor at the request of the Allies, who have since let her down and given her no assistance, because she chose, in the full plentitude of popular right, to recall her King.

Two states of unequal importance and discordant character will stand almost wholly beyond the influence of the Washington Conference. These are Russia and Turkey. The picture that we make of both is not a pleasant one ; but in reference to armaments they cannot be excluded, because the existence of their armed forces is primarily the cause of countervailing armies in the countries round them. If Poland, Rumania, and Greece are more immediately affected for the moment, it must not be forgotten how far Russia extends, or how insidiously the Turks are able to work upon Mohammedan sentiment in Asia and Africa. Nothing final in the nature of reduction in armaments can be settled until these two contumacious peoples rejoin the comity of nations, No one can say when they will. Neither seems to possess the capacity, either for evolution or for repentance.

There are also alliances, supplemented by military agreements, between certain states of Europe, which may tell against the conclusion of agreements to disarm. France has a treaty and a military agreement with Belgium, and, perhaps, understandings. at least, with other states. In the east of Europe the Little Entente unites Rumania, Czechoslovakia, Yugoslavia, and Poland in a series of alliances which Greece may possibly join ; and all these states may plead, not only these understandings, but their fear of warlike neighours, as reasons for maintaining their military strength.

For all these reasons we cannot be sure that disarmament, or reduction of armaments, so far as they relate to land forces, will have more than a *succès d'estime* at Washington. It is not a favourable moment to discuss this question, and it is even open to argument whether a direct attack on armaments is the best way of securing either their elimination or their abolition. I happened to take an unimportant part in the first Peace Conference at The Hague in the year 1899, when all the states of the world were not separated by the terrible antagonisms aroused by the late war. We were very well intentioned, very friendly, and set out to discover a formula for the reduction of armaments, in response to the late Tsar's humanitarian appeal. We could not find one, though we sought high and low for it, and a very good American delegation helped us in our search. Time has passed, and the urgency of the question may lead to the discovery of the formula for which we sought in vain ; but I am not confident that it will.

Recently I had the pleasure of meeting again that very competent Belgian lawyer, M. Rolin Jacquemyns, who also was at the Peace Conference of 1899, and is now the Belgian representative on the Rhineland High Commission. We compared notes and were both convinced that the creation of the Permanent Court of Arbitration at The Hague, which was the *chef-d'œuvre* of our Conference, was of more value than the League of Nations is ever likely to be. The Court still exists, and has done much useful work. To it should have been submitted the Upper Silesia case.

The Hague Court represents the main idea that seemed to me to be in President Harding's mind at the time of the late presidential election in the United States ; and I hoped that we were on the right track once more and were getting back to practical politics after our Geneva day-dreams. I shall retain that hope till the end.

An International Court of Arbitration, rather than a spurious form of world-government like the League, is the real remedy for most of the present troubles of the world. But I would like to see its importance magnified a hundred times by the acceptance of the principle of obligatory arbitration by all the states of the world. That condition we could not secure in 1899, because several states insisted on withdrawing from the purview of the

Court all questions in which 'honour and vital interests' were involved. That reservation practically made the Court useless at the time when it would have been most needed.

If the United States were ever great enough and wise enough to accept the principle of compulsory arbitration, I cannot name the state that would not follow her. Can any arbitral decision, even against the claims of any one of us, cause one millionth part of the ruin and loss of life and treasure of the late war? And, on the other hand, compulsory arbitration is a sure means of sterilising armaments, since, once international arbitration becomes our settled rule in diplomacy, the use of force must end; for no state would be so foolish as to keep up expensive forces for long when there was no use for them. On these lines, and I believe on these lines only, can the design that must stand behind the assembly of the Washington Conference be carried out to its logical completion.

I suppose that we shall not hear very much of the League of Nations at Washington. It was mainly American handiwork, but America's refusal to recognise her own child has relegated it to the political workhouse. No world-authority can exist when the United States, Germany, and Russia have no share in it. There are League enthusiasts here, as there doubtless are in America, and we must admire the devotion with which the League works and accumulates mountains of documents and reports. But we must also admit that it makes little progress and has scant authority.

Some say that the Council of the League is a mere creature of the French and British Foreign Offices. Others declare Geneva to be a focus of international intrigue. In any case, it is common ground that the League has no authority, and no force at its back except that of moral persuasion; and that it can do nothing but report, warn, or recommend. With difficulty it has at last agreed that the election of judges to an International Court of Justice shall be placed on its agenda at its second assembly, which is taking place as I write ; but I do not know why this Court should be any better than, or even so good as, our Hague Court of the first Peace Conference.

To take two years to begin to duplicate the machinery that we finished twenty-two years ago does not strike me as an achievement of great merit. The real practical international diplomacy of the moment, in all but American affairs, is controlled by the Supreme Council and by the Council of Ambassadors in Paris, both of which are, in effect, instruments for registering the decision of the Allied cabinets.

The League is left to its pious aspirations, and the main stream of diplomacy passes it by. Even when it has taken up a question like that of Armenia, with passionate earnestness, the only result has been that its protégé has become either Kemalist or Red ; while in the matter of mandates the United States has protested against decisions made without its approval, and the whole question is consequently hung up. Well may a French statesmen have said to himself sarcastically every morning in the spring of 1919, as he rose from his bed : 'Georges Clemenceau, you believe in the League of Nations.'

The Sorrows of Europe

In what particular manner President Harding and Mr. Hughes will change the situation for the better, we shall all learn presently ; but that the old Continent of Europe is beset with immense difficulties, political, social, economic, and commercial, is manifest to a traveller in every country that he visits. I place the question of exchange first among the anxieties of Europe ; and it is needless to remark how gravely British and American trade have been affected by it. It is not only the depreciation that has hit the world so hard, but the constant fluctuations,which have ruined confidence, caused every trader to think many times before he closes a deal, and involved, not only foreign merchants, but many British and American ones as well, in very severe losses. The foreigner, except in the case of a few neutrals, cannot afford to buy from us at the present rates, and consequently purchases only what he cannot produce or buy elsewhere.

In many cases, foreigners refuse to pay for our goods on arrival, because the local exchange has fallen since the order was given. In some cases, notably in Rumania, the inefficiency and inadequacy of the railway service preclude the forwarding of our goods from ports when they are landed ; and there the goods remain for months, on the quays, often perishing from exposure.

Is there no remedy against this deadly injury of the depreciated European exchanges ? I know of none except work, thrift, retrenchment, and time. But I think that we should explore the repudiation of old currencies, the replacement of old units by new, and currency reform based on the international redistribution of gold. Sound currency stands at the base of sound trade ; but as America holds most of the gold of the world, it is up to her to initiate reform.

People curse Versailles for not having stabilised exchanges at the time of the Peace Conference ; but when one looks into the procedure recommended, it is usually evident that the remedy is to declare that one crown, mark, franc, dinar, or lewa, is worth five, or possibly ten. Artificial stabilisation is financial quack medicine. International finance may be very clever, but apparently it is disarmed in presence of conditions with which it has had no previous acquaintance. Some people think, seeing how the hard-working countries like Germany undersell us owing to their depreciated exchanges, that their governments promote this depreciation. I have seen no evidence of it. The fall makes it enormously more difficult for countries to pay their foreign debts ; and those countries at all dependent on foreign imports naturally have to pay through the nose for them.

The depreciation, or, at least, the fluctuations, may be in part accounted for by speculation and gambling, which proceed on a vast scale ; but, taking the situation as a whole, the fall seems generally justified by foreign debts, by inflation, by internal exhaustion, by reduced output per man per day, by consequent failure of productivity, and by the inability of many countries to complete the reconstruction of their state machinery, without which their wealth cannot be fully exploited.

The countries doing best are those in which Labour is most moderate in the standards of wages and living it accepts, and in which governments provide cheap coal and relatively cheap food. This is Germany's strength. She is resolutely setting to work, and all classes are accepting a standard of living and of wages far below ours and even farther below the American scale. Compare the seventeen shillings per ton for German coal at the Ruhr pit-heads with the price we have to pay ; and compare the fifty pounds a year of the German bank-clerk with the pay of the English or American clerk ! This difference runs through all German social and industrial life, and there is, besides, a rigid elimination of waste, which is unknown with us.

The combination of the benefit from a depreciated exchange and that derived from low wages and poor living is enough to account for our difficulty in competing with German trade.

In many other countries the scale of remuneration of the highest dignitaries is preposterously small. In Austria the President of the Republic draws only eighty pounds a year, and heads of departments in the Foreign Office tell me that they cannot afford a new suit of clothes.

The High Court Judge in Bucharest draws sixteen pounds a month, and the lieutenant four pounds. How they manage to live at all, with prices at their present height in these countries, is one of those mysteries which I have not been able to penetrate, though we must, of course, admit that the purchasing power of the local currency in the country itself is much higher than the English or American equivalent of it would in London or New

York. A few countries have checked inflation and are bravely facing their liabilities ; but in many—and Poland and Austria are the worst cases—inflation goes on, and selfishness often prevents the imposition of taxes needed for reconstruction.

Generally speaking, I regard this question of the rates of exchange as much more vital to England and America than to Continental Europe, though in one way or another all suffer from the present situation. We are really in presence of a state of chaos which injures all the world, and only the union of the world for the purpose of mending matters can improve conditions. In this matter, America might take the lead, and, by collecting the best practical experts, endeavour to formulate a solution. The Brussels Economic Conference gave us the most excellent advice upon the questions of state finance and economics ; but something more is needed before we can go ahead. Unless some financial genius can discover a remedy, one must regard British and American trade with Continental Europe as almost dead for a long time to come.

Second only to the exchanges, there comes the urgent need of freeing international trade by every possible means from the very great obstacles which are at present accumulated in its path. I refer especially to pasports, custom-houses, tariffs, permits, and all the vast machinery for selfish national isolation which seems especially devised, not to assist trade, but to hamper it

The grand tour of Europe is no joke in these days. One's passport becomes a formidable document. One must get a visa in advance for every country through which one passes, even if one does not propose to stop there. One must carry only a very limited amount of the local money out of each country ; and in travelling across a number of states one must carry the coinage, or rather the horrible paper, of each. The trader is greatly handicapped by a system of permits, and export and import duties, and the wonder is how any trader gets a ton of goods into, or out of, any country. This arises from state control of trade, and everything shows that, whatever else the state may be, it is a failure as a merchant.

We see the system at work to kill trade in full perfection in the Succession States of Austria. The old Austro-Hungarian Empire was favourably situated economically, because different parts of it supplied things that other parts lacked, and everything passed freely from one province to another. There was internal free trade, and the Empire was almost self-supporting. Hungary sent her wheat and her timber, Bohemia sent her coal and sugar, Styria and the other parts all their products. It was less the Austrian marriages that made Austria happy than the very shrewd business sense which realised that certain provinces were needed to supply Austria's deficiencies.

Now all this economically happy state of affairs has terminated. The Succession States have all closed their frontiers against Austria and against each other. Each has its own currency, and has set to work to build up customs barriers on every side against the territories with which it once traded freely. This has injured the present Austria most, and has indeed reduced her state finance almost to extremities by compelling her to pay vast sums for wheat and coal. But before long the selfish Succession States found that, in injuring Austria, they were losing their customers and injuring themselves ; so, by the natural force of circumstances, we shall in due course see a change of policy for which Austria, Hungary, and even Czechoslovakia are almost ripe.

But the big idea of Dr. Benes, the Foreign Minister of Czechoslovakia, to create the United States of Central Europe by a series of tariff agreements between half a dozen states in this part of the world, may take long to be carried out ; more for in some quarters the tendency is still to pile on duties, chiefly in order to collect money, but also to protect home industries.

The broad fact remains that international trade is grievously hampered, and that it should be our object to free it from its fetters, both for our own sakes and for the sake of these small countries which are busy strangling each other to no possible benefit for themselves. I believe that the quickest and most drastic cure for the evils of Europe, and failing currency and exchange reform, would be a year of completely free trade, with no tariffs at all, inward or outward; but one must confess that the nations concerned, not to speak of others, have not yet reached such a state of grace as to accept a remedy of so novel and so violent a kind. The tendencies, on the whole, are the other way. Even on the international rivers, the smaller riverain states are most tenacious of what they call their rights, and claim powers which the régime of international law does not allow them.

All governments want money, whether to administer the state or to reward political friends. Therefore the rule is to tax everybody and everything, but especially the foreigner. The export duty on Rumanian oil is a typical case; for, if it hits directly the foreign capital invested in this industry, it also injures a source of local wealth, and gives a subsidy to other states which supply oil. The idea of a fixed export tax, laid on regardless of world-prices and falling values, is one which must have originated in a lunatic asylum.

In other places we discover a consortium, or government trading-machine, which supplies posts for political adherents, usually ignorant of trade needs and practices; and it need scarcely be said that it trades badly and imposes on the produce of the country quite needless losses, often failing to find markets at all. In short, there is every grade of incompetence to be found as we pursue our inquiry; while, of course, the immense loss and damage of the war has thrown numerous states into a disorganised condition and communications have particularly suffered.

Another change, which we in England, at all events, watch with some anxiety, is the agrarian policy, which has taken the form, in several states, of distributing the land among the peasants. It may have been, and it was in some cases, a political necessity, and may have prevented an agrarian revolution ; but the effect which it will have upon the export of cereals is of considerable interest to the world. The great estates are being broken up and replaced by small holdings, which usually run from some three acres in Alpine regions up to twenty acres in average arable land, rising again to six hundred acres at most for the old proprietors.

There is no universal scale, nor even the same scale in all the provinces of each separate country ; but the general effect is to replace large landed properties by small ones, with various scales of compensation—all very low—to the former landlords. Most of these laws were passed in the first rush of revolutionary enthusiasm after the war. In some cases they have been widely applied, in some partially, and in others scarcely at all. But all the laws stand, and it is the general belief that the exportable surplus of cereals, and especially of wheat, will diminish with a generalised peasant-proprietorship.

The tendency of the small holder is to grow patchy crops, primarily for his own food and that of his family ; and there will not be the capital necessary for rich manuring, for providing modern agricultural machinery, or for purchasing high-class stock. On the other hand, a plurality of landowners means more stable political conditions, and may lead, some hope, to increased production, owing to the personal interest of each small farmer in his land.

Some attempts have been made by the proletariat, notably in North Italy, to seize factories and to exploit them for the exclusive benefit of the workers. These attempts have failed, because the new men in possession found themselves quite incapable of managing the administrative part of the work, the contracts, and the sales. They, therefore, in many cases, invited the old proprietors and managers to return, while the bourgeois parties created the *fascisti* in Italy, and took other measures to defend themselves.

In general, the tyranny, the excesses, and the fearful results of the Russian Revolution have sunk deeply into the minds of the workers in Europe. If Bolshevism had been specially designed to expose the futility and uneconomic absurdity of the theories of Karl Marx, it could not have more appropriately carried out its mission than it has done during the last four years. The error, and the tragedy of the error, have been denounced to the workers of Europe by many missions to Russia composed of men of extreme views. With few exceptions these men have confessed themselves horror-stricken by the conditions they have found ; and, though Communism is not everywhere dead in Europe, there has been a powerful reaction against the disruptive theories of a few years ago. The affair really came to a head in the Bolshevist invasion of Poland ; and if the failure of that attack did not convince Lenin and his dupes of the futility of their theories, it conveyed to them, at all events, a sense of their weakness against even partially trained troopers ; and since then Bolshevism has been steadily losing ground in countries other than Russia.

There are some Communistic centres in Europe where outbreaks of this disease may recur, but I do not know the country in Europe which has any serious fear now that its people can be stampeded by the fanatics of Moscow. The experiences of Berlin and Munich, Vienna and Budapest, have sufficed. The country has one hold over the towns : it can always starve them.

The disruption of four great historic empires, and the substitution for them of various forms of democratic rule, have naturally caused immense disturbance in the political atmosphere, and the political weather is most uncertain. Bulgaria keeps her dynasty, and Austria thinks more of joining Germany than of recalling the Hapsburgs ; but Hungary is monarchical, and would have a king to-morrow if she dared ; while a large and influential part of the German population remains in principle monarchical, and desires to revert to that form of government. The German Empire acquired its former great position under a kaiser, and every German is regretful of the past.

The present government of Dr. Wirth and the personality of this honest Swabian, are very highly esteemed by the Allied and associated diplomatic bodies in Berlin. Chancellor Wirth is endeavouring to do his duty by the Treaty of Versailles, as well as by his own people. But he has to call upon the German people to double the state revenue in order to pay reparations; and though I am convinced that he can do it if he meets with proper support, politics in Germany are very bitter, and the parties of reaction stick at nothing. All the old reactionary forces are still in existence. The Army, the Church, and the Universities combine with the landlords and the great industrial magnates to make things difficult for a government which has no great prestige for want of past successes and has the invidious task of sending the hat round for the Allies.

The mass of the Left, and even some of the intermediate parties, have at present rallied to the Chancellor's support; and, if street demonstrations count for much, the majority of the voters are for him. The Allies have abolished the Rhine customs as a tribute to him; but, owing to the opposition of France, have not withdrawn from Düsseldorf, Ruhrort, and Duisburg, as Dr. Wirth has very earnestly pleaded that they should.

The Right parties in Germany complain that the Government lacks authority, cannot represent the country with the old distinction, and is subservient to the Allies. Most of the notable leaders of the Imperialist party are getting on in years, and they probably feel that time is on the side of German Republicanism. In a few years most of the old officers will have settled down to new occupations and may retain little more than a sentimental attachment to Kaiserism. The Right probably feel that they cannot afford to wait, and they count, with some reason, upon the national pride, which revolts against the peace and the surrender to the Allied ultimatum of last May. But it seems to be the prescriptive right of this party to make colossal blunders, and the assassination of Erzberger, almost condoned as it was by many Opposition newspapers, is the last on the list.

No one can safely predict the future of German politics, which depend on events that cannot be foreseen ; but that the character of the new Chancellor and the policy of his Government offer the best ascertainable chance for the gradual pacification, not only of Germany, but of Europe, will not be disputed by the closest observers of European politics.

For the reasons stated in the earlier part of this article, I do not think that very much can be expected from the meeting at Washington in the way of reduction of land forces. With respect to navies it is different, because there are only three great navies that count, and none of these is specially concerned in the enforcement of the terms of peace upon our late enemies, who have no navies at all. It is, therefore, merely a question of agreeing to a mutual standstill in naval armaments ; and this question, it would seem, should present no insuperable difficulties.

But I cannot think that such an important conference will break up without suggesting a remedy for the ills which I have briefly described. Armaments are symptoms of a political disease, but are not the disease itself. The real diseases of the world are unstable exchanges, unsound currencies, hampered trade, and the false nationalism which shuns obligatory arbitration. Cure these diseases and armaments cure themselves.

CHAPTER IV

The Great Middle West

I. Men and Women

A Lecturing Tour—The American Love of Lectures—Attentive and Friendly Audiences—Generous and Warm-hearted People—American Men and Women—Absence of Loafers—Business the Play of America—Great Openings—American Women and Literature—American Politics—Lack of Leadership—Export Trade—Banking and Business—Gold Accumulations—Political Changes—The St. Lawrence Waterway—Importance for American and Canadian Wheat Exports—Hostile Influences—Travel Westwards—An Untidy Country—Optimism—A Land of Equality and Equal Opportunity—Cost of Living, Rents, and Wages All High—No Fixed Limits for Wages or Hours of Work—Research, Business Organisation, Distribution, and Sales—Resulting Economic Mass Production—American Individualism—A Real Motive for Life and Industry—The Plain People—A Fair People and a Religious People—TheRotarians—English Speakers—The Press.

A visit to Washington whets one's appetite for travelling in America. One finds at Washington many interesting people and many interesting ideas ; kind, charming, hospitable people, and ideas which take one out of the European ruck, but one is afflicted all the time, as one is at New York, by the sense that one is living in a world of its own that is not completely American. There are too many Irish, Poles, Russians, Germans, and Jews in New York to allow us to feel ourselves in a distinctly American environment, while at Washington there is something of the starched feeling belonging to every official world. One longs to get into a large car, and, after travelling a few hundred miles, to get out and go up to the nearest people and ask them what they are doing, and thinking, and feeling.

So, soon after the Washington Conference, I went off to the Middle West to try to find out what sort of people the real Americans were after all, for somewhere in this vast territory I imagined thatI should find plenty of them and have a glimpse of their minds. It was no good going there without some ostensible object, because one would not in that way meet enough people or be brought into close contact with all the grades and gradations of folk. So I invented a few subjects for lectures, and that gave me a sort of fictitious status. I chanced to hit upon two subjects which the people happened to take an interest in, and they would never let me talk of anything else. These subjects were the State of Europe and the Washington Conference. The latter enabled me to feel the pulse of Americans about their own affairs, and the former showed me the state of feeling in America about the old Continent which seems, when one is in America, to be as far removed from one as a distant planet. So I moved about from St. Louis to Boston and from Buffalo to the South, visiting numerous towns, small and great, talking in theatres, town halls, churches, conventicles, at city clubs, ladies' clubs, university clubs, and so forth, as well as at private houses, rotary conventions, twentieth century clubs and such charming meeting-places as the New Arts Club, where so many English lecturers have been given a send-off on their careers of lecturing crime.

I do not suppose that there is any better or quicker method of getting to know the people of America. They have a playful habit, especially at the city clubs, of asking one a series of very pertinent questions when one has done talking. It is the best of the fun. They ask all sorts of questions. What do you think about the American Senate? What is your view about the Panama Canal tolls? Is the British Government frightened of Mr. Gandi? And so forth. But one soon learns that if one can turn the laugh against the questioner one has one's audience with one. The Americans have the habit of attending lectures. They like to study everyone whose name they know. Busy people themselves, they have scant leisure for travel or study, and like to be informed. They spoil one for any other audience. They ask one to speak for an hour, an hour and a half, or two hours. They never seem to tire. They are the most courteous, attentive, and friendly of audiences, and their questions usually display a lively interest in the world's affairs.

No people could be more hospitable, more anxious to enable one to see everything in or near their town, and more eager to help one in every possible way. One must, on these tours, be immune from fatigue. One must do with very little sleep. Everywhere there are people to meet one, mostly perfect strangers, but as kindly as if they had known one all one's life. One's time is mapped out by them, and one must place oneself in their hands. It is worth doing. The memory comes back to us afterwards of a thousand acts of kindness and thoughtfulness, amongst all sorts of men and women, that can never be repaid. But they certainly do make us exceedingly grateful to them, and there is real regret in the parting from new friends so kindly, so generous, and so warm-hearted.

We probably all know a number of rich Americans, men and women, who spend much of their time in London, Paris, and the Riviera, and have learnt the gentle art of sauntering, having nothing to do. It is with astonishment that one finds no such people in America. The fact seems to be that if one wants to loaf one must go to Europe, for in America there is no one to loaf with. It is considered almost a disgrace in America to have nothing to do. All the men, rich or poor, work all day. Business is the play of America. Success in business is the universal aim, not money-making, though it sometimes comes to the same thing. But there is a distinction, and one must realise it. There are few of our agreeable sports and pastimes in America. Golf has come to stay, but it only affects a very limited class. There are few occupations for idle men, and it is not wonderful when there is none idle. Even the millionaires work as hard as the rest, and frequently die in harness, though able to afford themselves every luxury in the world and to live where they please. The son of the millionaire, if he does not leave his country to loaf abroad, takes up some, almost any, profession, and works at it as hard as he can. No one seems to mind what business it is.

There have been great prizes in American life. There are fewer now. But every real hard worker can still make his pile. The country is so big, openings are still so numerous, there is so much scope for brains, activity, and intelligence, and there is still such a youthful enthusiasm prevailing everywhere, that the future seems bright, and optimism refuses to surrender its rights even in the most depressing circumstances. So all the men work, but one of the results is that there is no leisured class to be found anywhere, and to this fact one must attribute the result that America, for her population and intelligence, has not given to art and letters, to erudition and culture, anything like her proper share. The man goes to work early. Luncheon is not a meal, but a particular aspect of the day's work, for it usually takes the form of a talk with business men. So in the evening, when a man comes home, he has had quite enough for the day, and few turn to books.

It may, and very probably will, be that women will take a leading part in art and letters in America. I noticed the last month when I was there no fewer than five articles by ladies in the leading American review. Ladies' clubs play an increasingly notable part in American life, and in the great cities they include all the best known women in society. I cannot explain what is meant by the word 'society' in most of the large American towns. I do not know, and cannot imagine, who gets into it or how. It is one of the mysteries. Seemingly, in the minor towns, great success in business lifts one out of one of the strata of life and places one in a higher. Then one lives with one's successful friends in a set of some fifty people or so, and apparently one has to desert one's old friends in the process. This is so antagonistic to the natural American that I suppose I must be wrong. It is a particular feature of American life that one must be an American to comprehend. I give it up. I am only sure that the women of America are the only people with spare time on their hands, and if this goes on I imagine that they will have much to say in the development of American literature hereafter.

II.—Politics and Business

The home politics of America are fearsome, and require to be approached with precaution and to be treated with awe. It is possible that 999 people out of every 1,000 in England are not aware that members of the American Government have no seats in Congress, and are consequently unable either to produce their measures or defend them, or answer questions, in the national Parliament. They can be hauled up before committees and examined, but the initiative is not with them, so that the executive and legislative powers are not only separated, but divorced, and the judicature is equally alien to the other two.

This was the intention of the original Constitution, and it has not been modified to any appreciable extent. President Wilson, indeed, revived a practice which had not been used for a century before his day, and hit upon the plan of not only sending a message to Congress but of taking it there himself and reading it. It was an idea which, I am told, Colonel Roosevelt had never thought of, or he would certainly have employed it. President Harding has continued it. It is a link, but from our English point of view a very inadequate link, between the Executive and Congress. Imagine our Parliament deprived of its front bench and all its other powers of questioning, discussing, debating, and attacking Government measures in the presence of Ministers, and then picture the mighty difference between the two conceptions! It is doubtful whether this American system can last very long. It is out of the picture of modern Parliaments, and does not find many people to defend it.

The framers of the constitution certainly foresaw the dislike of English-speaking people to be governed by anybody for long when the quadrennial Presidential election was introduced. But the constant changes, not only in the Presidency, but in the Senate and the House of Representatives, owing to the frequent renewal of Senators and Members, and the upheaval which affects all the Administration in case of change of Presidents, tend to make politics unpopular, and the careers of diplomatists in the higher grades a matter of chance. Except the leading men of the hour, few American politicians are known outside the United States, and a good many are only known within their own individual States. There is, strictly speaking, no assured political career, and consequently the best men very rarely follow politics. There results a real lack of leadership, while the member of the Executive rarely stump the country to spread the ideas of the Government far and wide. The size of the country tends to prevent this course, and so the political platform, except during Presidential or State elections, is replaced by the Press, which becomes a very considerable power in the land. Such papers as the *New York Times* and *World* reach Chicago about noon the day after publication but, generally speaking each great city has its own Press, usually under independent direction, though there are a few groups of papers which obey a central *mot d'ordre*.

With two parties, Republicans and Democrats, votes tend to equalise themselves, or at least to make a two-thirds majority exceedingly difficult to secure, whether in Congress as a whole for the purpose of amending the Constitution, or in the Senate for ratifying or rejecting a Treaty made by the President with foreign Powers. Even apart from the justice of the cause, an American Government has not much inducement to submit a treaty to the Senate, since the prestige of the Administration is diminished by a hostile vote, and it must be added that the country, as a whole, does not favour foreign treaties or foreign entanglements of any sort.

One feels instinctively the reason when one gets so far as the Mississippi Valley. The country appears to extend, with indeterminate limits, in every direction. Europe appears to us very small and very distant, and to be troubled and troublesome beyond all sense and reason. The United States is more or less self-contained, and can get along with its own internal trade. Agriculture is a huge business, and there is coal and oil and every description of raw resource and industry. Why bother about the foreigner? The foreign trade of America is only from 10 to 14 per cent. of her total trade, and exports have not the same importance for America as for us. But though this may be true, America has felt very severely the loss of the purchasing power of her European clients during the last two years. The farmers have begun to understand that they are deeply interested in the welfare of their foreign clients, or, at least, the intelligent people among them, and those in charge of their various farmers' associations in particular. Similarly, the value of exports to Europe fell by about three billion dollars in 1921, and was particularly sensible to all business houses. These things have given the splendid isolation parties a good deal to think about.

It has been the same with the bankers. Gold has flowed steadily into America since she came into the war, until she finds herself with huge reserves of gold, far exceeding all the needs of her currency, and producing no return at all. So the Federal Reserve Board came to the conclusion that the United States had an interest in the introduction of some form of gold standard as a means for the resumption of the old trade relationships through the effective stabilisation of exchange, and even went so far as to affirm that the present abnormal concentration of gold in America was a source of danger, because it was a false guide in matters of credit policy, and no longer an index of the outside limit of legitimate credit expansion. I have ventured to tell American audiences that what they have stored in their strong-rooms is not gold at all, but the purchasing power of their European clients, and, so far as I can see, America is prepared to discuss terms for the redistribution of her accumulated gold. But we must not think the project simple. There are involved many complicated problems. Reparations and international debts will have to be regulated when America lends her aid, and the effect of national policies upon the amount and the direction of current expenditure will also have to be explained before the American safes will be unlocked.

America does not very much trust Europe. She is not very well posted in international finance. She is shy of the assumed astuteness of foreign statesmen, and looks somewhat askance at the international banker. Yet the conditions in the United States are such that in her own clear interest she is bound to join hands with us sooner or later. It is becoming obvious to everyone that the world is one great business establishment, and that there is no Power, however rich, self-contained, or strong, that can afford to stand apart and take no interest in the reconstruction of Europe and the semi-Asiatic communities abutting upon it. Least of all can a great producing country like America, which developed immense power of production during the war, and can hope for no serious return of prosperity until Europe can again afford to buy American goods.

Are there any great changes visible on the horizon of American politics? Yes, there are some. There is a tendency amongst certain groups of interests to place these interests above those of the party ticket, and to see which candidate will satisfy them before giving in their votes. By these means a strong and resolute minority can carry a policy through without being actually represented in Congress, though it may have a majority there pledged to support it. The farm bloc is the most powerful of these groups, and others are likely to copy them. The difficulties of the administration over the soldiers' bonus are an example.

An American Government—of the people, for the people, and

by the people—keeps its ear to the ground to catch the first rumblings of distant criticism, and is rather the servant of public opinion than its director. There results a somewhat feeble action in Congress. The return of Mr. Albert J. Beveridge in the Indiana Republican primaries by 20,000 votes over President Harding's close friend, Senator New, so far as it was not a personal triumph for a man of great capacity and intellectual force, was a protest rather against the invertebracy of Congress than against the President himself. The old party ties still hold. Neither Labour nor the farmers yet venture to contest them. But Mr. Beveridge's success was a protest against Congress shuffling, and a warning that unless more of the Roosevelt spirit inspires Congress there will be trouble in the future.

III.—The St. Lawrence Waterway

Many great schemes are hatching in the busy brains of the Middle West, but there is none to compare with the proposed improvement of the St. Lawrence River, between Montreal and Lake Ontario, to enable ocean-going ships to serve the American and Canadian harbours in the region of the Great Lakes, and to carry cargoes to and from this region without breaking bulk.

The deeply interesting report of the International Commission, which had considered this subject, was laid before Congress by President Harding, who favoured the scheme, as did some eighteen States of the Union, as well as Ontario, Alberta, Manitoba, and Saskatchewan on the Canadian side. But there are lions in the path. New York, both State and City, are in opposition. So are Montreal and Quebec in Canada. The railway interests, on the whole, are unfavourable. So are important bodies at Buffalo, Boston, Philadelphia, Providence, and Troy. It is the Middle West against the East, but there is greater weight of numbers among the supporters of the waterway than among its opponents.

The idea that the vast products of the American and Canadian Middle West should be shipped from the many excellent ports of the Lakes and be landed at Liverpool without the cost, delays, and damage of transfer is certainly enticing, even if open water cannot at present be counted on for more than seven and a half months of the year. Similarly, there would be no break of bulk on British and other goods sent to the Lake region, which is a great consuming, as well as producing, area.

The Commission see no good reason why ocean-going vessels of suitable draft should not safely navigate the waters in question, as well as the entire waterway from the Gulf of St. Lawrence to the head of the Great Lakes ; while further they believe that there will be such a volume of trade that it will justify all the cost of the scheme. One cannot expect anything else. In 1870 there passed through the canals of Sault Ste. Marie some 49,700 bushels of wheat and 304,000 bushels of other grain. These figures had increased in 1920 to 255,481,558 and 51,630,135 respectively, and as the limit of the grain-growing capacity of the States in the American West has not yet been reached, while the output of the Canadian prairie provinces is in its infancy, the next ten or twenty years may see grain shipments many times greater than the 1920 figures.

The population and output of the American side of what is termed the "tributary area" of the Lakes naturally exceed those on the Canadian side. On the former there was a population of 42,000,000 in 1920 to the 5,000,000 of Canada. On the American side, 63·3 per cent. of the aggregate value of all farm property in the United States is within the zone of the Lakes, and its value is estimated at 50 billion dollars, as against the 78 billions for the whole of the Union. This great wealth of the American Middle West is due to a large and industrious population, fertile soil, and adequate rainfall. On the Canadian side the exports of wheat averaged 131,500,000 bushels over the years 1916-1920. As there is expected to be a saving of 10 cents a bushel in cost of transport if ocean-going transport is used, the potential advantage of the scheme to the farmers and their clients overseas is clearly manifest.

One of the main reasons for the strong demand by America for the new waterway has been the inadequacy of railway transport facilities. The inability of the railways to handle the traffic expeditiously has resulted in great delays and heavy losses. It is declared that the railways must spend 15 billion dollars in the next ten years, in order to equip themselves to handle the traffic efficiently. As there seems little chance of any such expenditure, the need for the new waterway becomes all the more crying. These conditions are not repeated on the Canadian side. In the same way, there are not the same complaints about the railway terminals in Canada that there are in America. But there are greater possibilities of expansion on the Canadian side, while the power plant involved in the plans is almost sure to result in a great growth of manufacturing industries on both sides of the border.

On both sides there is a desire, on account of rapidly-expanding business, to utilise every practicable means of communication; and the great natural waterway of the St. Lawrence, leading as it does into the very heart of the Middle West, cannot be neglected without risk. The total cost of the scheme proposed by the Commission is $252,000,000, and it is very fairly proposed to divide the cost between America and Canada in proportion to the benefits which each receives from it. This ratio will obviously require periodical revision, but the cost of the scheme appears to account for the lack of enthusiasm of the present Canadian Government in the enterprise.

It is estimated by the experts, Mr. Roy S. MacElwee and Mr. Alfred H. Ritter, that the scheme will develop 4,000,000 h.p., worth about $60,000,000. Such power, if used for manufacturing, would produce good returns.

Water power is regarded as subsidiary to navigation in the scheme. The details of this scheme are scarcely worth description at present, because, athough an engineering board of two officers has sent in a report, this only deals with outline plans and lump-sum estimates. The Commission, while accepting the report, recommend that it should be referred back to the board which, they suggest, should be enlarged by other leading members of the engineering profession, so that the whole question may be more thoroughly studied. The Commission approve of a scheme of dams, side canals, and locks, and propose a depth of 25 ft. in the canals and 30 ft. on the sills of the locks. A study of Lloyd's Register and of the traffic passing through the Panama and Suez Canals shows that these measurements very fairly meet the needs of the present time, and the average draft of 85 per cent. of sea-going steamships.

It will be seen that the whole plan is full of interest. It deals with large questions, some of them contentious, all of magnitude and complexity, and, as a whole, involving national and international interests of very great moment. The history of the waterway, the physical characteristics of the great basin of the St. Lawrence, the existing and proposed canals, whether main or subsidiary, the calculation not only of grain supplies, but of iron ore, coal, lumber, and so forth, the study of railways versus waterways, and of time and cost, questions of ships, drafts, and harbours, insurance, refrigeration, ice conditions, and many other matters, all enter into the problem, and each deserves longer treatment than has been given to the whole scheme in this short chapter. There is also the question of the Georgian Bay Ship Canal project which has many advocates in Canada, but at present, as the St. Lawrence scheme stands approved in principle by the American-Canadian International Joint Commission, it holds the field, and some decision must be reached upon it before alternative schemes can be taken up.

The whole question is one for the Canadian and American Governments to settle between them, and a very reserved attitude of impartiality is necessary on our part. But it would be absurd to pretend that we are not deeply interested in seeing the general scheme carried out in one fashion or another. The loss of the Russian wheat has been a heavy blow to Great Britain, and no one can say when this market will be reopened. In the American Middle West and in the prairie provinces of Canada is the largest existing surplus stock of wheat and other grain available for export, and the arrival of this grain at our ports in the shortest possible time, with the least possible loss and damage, and at the cheapest possible price, is naturally of deep interest to our consumers. It is probable that some three-fourths of the Middle West wheat harvest can be sent across the Atlantic before the waterway is held up each year by ice, and in this case the American and Canadian farmers and their bankers will be very greatly relieved and benefited.

When Montreal and New York State and City are united in opposing a project of this sort, and when Buffalo and Quebec, Rochester, and other eastern towns unite with them, we are certainly bound to hear, and we have, indeed, already heard, everything against the scheme that can be said. All these places profit by the present arrangements, and will be the losers by the change. But local patriotism gives way in large affairs to national patriotism, and the future agricultural and industrial expansion of the Middle West is so certain that no alternative means of communicating with the outside world can safely be combated. The present may not be an ideal moment for the provision of the large sum required by the scheme, but the advantages, both to the farmers from the ocean transport and to manufacturers from the power scheme, promise returns which seem to justify the optimism of the Commission.

IV.—Plain People

As one travels westward in America one becomes conscious that it has been traversed by some huge army without a salvage corps. It was the army of the pioneers, who took the cream off the country as they advanced, left an intolerable deal of rubbish behind, cut down the trees anywhere and anyhow to build their shanties, and then passed on seeking for other fields to conquer, leaving part of the army behind to make the best of things. A more untidy, unkempt, and unfinished country no one can name. It will take a century or two to be put in order and cleaned up.

Not much even of the architecture of the present day will remain, though the new architecture of some of the universities, theatres, museums, and churches is very fine indeed. As the Americans scrap their most expensive plant in their factories when they find another which can give them a 10 per cent. pull over their rivals, so they will scrap their towns and streets, and even some of those charming residential quarters where American individualism comes out in the fact that no two houses in these quarters are alike. There was never a progressive people so conservative, nor a conservative country so progressive.

The Americans are still very young, and have all the delightful optimism of youth. They believe in themselves, their country, and their resources. One does not meet pessimists in America. Why should one ? A man of brains and activity has all the world open to him, and no hide-bound unionism ties him down to the pace of slower and more stupid competitors. A land of equality and of equal opportunity, where intelligence is rewarded and can aspire to the highest prize. The cost of living is high and rents are high, but wages are high too, and there is no limit for the men who are both skilled and industrious. No fixed limit for wages and none for hours of work. The American is not a better workman than the Englishman, but his output is larger owing to his freedom from shackles, while the employer and the manager are quick, intrepid, enterprising, and alive to the virtues of research work. The art and the science of business organisation, sales, and distribution have been pushed to the extreme limits of possibility, resulting in economical mass production unrivalled elsewhere.

High wages and unrestricted output are twin brethren. The American employer knows that trade unionism has nearly stifled industry in Europe, and so he fights it to the death, so far successfully, and he is helped by the strong individualism of the men, who suffer no limits to be set to the product of their talents. Trade unionism in America has not the strength of ours. It may grow and may be better led, but at present it conflicts with the American idea of liberty, while the size of the country, the distances, the mixture of races, and the masses of immigrants who have hitherto flocked to the country, but are now being limited in their numbers, tell against the success of the trade union movement.

American individualism is not our English or Scottish individualism, with its feudal or clannish restrictions of caste and class. It is the exact antithesis of the Marxian idea and of the futilities of the Soviet tyranny. It is very broad and deep, this American individualism, and has been the secret of American achievement from first to last, whether in the realms of war or peaceful progress. It does not ban the principle of co-operation for mutual advantage by any means, but it must be association founded not on mass ideas, which are usually emotional and rarely sane, but on the free exercise of initiative and character by the best of the bunch, whatever the form of the association may be. There is a real motive for life in America, because the highest prize is open to talent. The marshal's baton is in truth carried in the pack of every American workman, and in no town that one visits can one fail to find men in high and responsible positions, including many of English birth, who have worked their way up by sheer force of character.

There is everywhere an intense desire to create, but to create one must first arrive. It is no good to arrive unless character and competence are fortified by strong morality and a clear sense of right and wrong. The American people, taken as a whole, are a fair people, a religious people, to whom nothing shabby or crooked can long appeal. Periodical outbursts of wrong-headedness are usually the result of ignorance and immaturity. One must never forget the stream of ignorant immigrants which has poured into America for so many years, and takes two or three generations to become merged in the common stock by learning English—rarely accomplished in a first generation—by education, and by gradual assimilation of American ideas.

One still finds great blocks in towns where the German, the Pole, the Russian, and the Czech live lives apart, and this natural segregation tends to the prolongation of racial differences. Maybe an employer can always break a strike or a nascent union by drawing upon these labour resources, but one also asks how he will get all the heavy and dirty work of business done when this source is dried up by limiting immigration. The Middle West is not the right place in which to study the negro question, but one fact stands out, namely, that the negroes are less concentrated in certain States than they were, and that their dissemination throughout the country tends to lessen, though it does not remove, the gravity of the problem which turns upon their future.

The Churches have done much for America. They form centres for the expression of thought, and nothing was more remarkable during the Washington Conference than their strong support of disarmament. There is a tendency, if not to the union of all the Churches except the Roman Catholic and the Unitarian, at least to their co-operation and interchange of ideas which are manifest in many ways. It is also not possible to ignore the great hold which the Rotary Clubs are beginning to exercise upon American thought. Neither religion per se, nor politics, have a part in the Rotarian system, which is founded on good fellowship and on the idea of service to the people. Here we find the first serious symptoms and aspirations of altruism which may hereafter modify the excesses of individualism and lead on to some higher and nobler expression of it, provided that politics are absolutely excluded from its purview.

In the great Middle West one meets countless numbers of the real plain people of America, who are at once the foundation, the stay, and the expression of the whole nation. A good people, men and women, lovers of right and justice, friendly and genial, they convey a more true sense of the meaning of liberty, fraternity, and equality than some European people who advertise these terms. They have no strikingly different views of European politics from those of the Eastern States, but, by reason of their greater distance from Europe they are less well-informed than New York and Washington of passing events. So they ask, not once but a thousand times, why more Englishmen do not come and talk to them and tell them what is passing in their minds. They want to know all that one can tell them of personalities and politics in Europe, the why and the how, and what it all means, whether it be of the Great Powers or the minor States, or of India and the British Dominions. They also like to get a British opinion upon American ideas and politics, in order to confirm their ideas or otherwise, and one need not trouble to draw conclusions, for Americans are quite capable of forming their own.

Men like Mr. Reginald McKenna, Senator Pearce (of Australia), and Major Beith (Ian Hay) have done unimaginable good by their speeches in the Middle West. As for Sir Auckland Geddes, though he has probably spoken to more American towns than any former Ambassador, there are towns which positively complain that he has not visited them. They seem to think that they have a prescriptive right to the British Ambassador's time.

One must certainly admit that liberty degenerated into license in many great towns of America of late, and that a wave of crime at one time swept over the country, to the despair of its friends. The cause is put down to the war, but the excuse seems inadequate. It is true that the many unemployed on the streets are partially the cause, and that the war, or rather its after effects in checking American production, is the cause of unemployment. But the main reason is rather the absence of suitable legislation against the carrying of arms and explosives and of punishments for the infraction of such laws.

The American Press has a positive genius for discovering sensations of every kind and for giving them unworthy prominence. But the Press in America is a business proposition. It gives the public what they want, and does not attempt to pose as the arbiter of public taste or morals. In very few instances are the editorial articles in the papers now worth reading. It is the special article written by men who are known that attracts readers, and the American magazines are very good indeed. Whatever the public want the Press gives. The public wanted to know every single detail of the wedding of Princess Mary. It was given columns and columns, and could not have enough. An analysis of the reasons for the extraordinary enthusiasm would be worth giving, but perhaps one may limit oneself to the statement that a Princess still reigns supreme in the American fairy tale, and that here was a real live English Princess in a golden coach who personified all the beautiful memories of nursery tales.

CHAPTER V

The American Military System

(See MAP)

Resemblance of American Military Organisation in the Past to Our Own—Before the World War—America Preserves Her War Material After 1918—The American National Defence Act of 1920—The Regular Army—The National Guard and Organised Reserves—Training Corps—Retention of the Old War Divisions—Six Armies of Fifty-four Divisions—Army and Army Corps Areas—The Reserve Officers—Observations—Principles of the New American Plan—No Compulsion in Peace Time—Comprehensiveness of the American System—The Air Service—Artillery—Tanks—Great Results Attained—1917-1918—Our Memories of American Co-operation in the World War.

THERE are two methods by which we can usefully examine a foreign military system, one by criticism of its faults and failings, and the other by an examination of the spirit which presides over the plan, and of the results which this plan would achieve were there no such things as parliaments and politicians. It is from the latter point of view that the new American Army system can be most usefully examined by us, for assuredly we should be the last people in the world to wish to throw stones at anybody in matters of military organisation. The military conservatory in which we live might all too readily lend itself to reprisals.

The history of army organisation in England and the United States has had a striking and almost fantastic resemblance, due partly to the origins of the American Army, and partly to the identity of political methods and military conceptions in the two countries. A few years before the World War the American Army much resembled ours of pre-Cardwell days, and when each of us set to work to create a national army we each had voluntary service as a basis, each a small Regular Army, and each the nucleus of a second line in our Territorial Force in Great Britain and in the National Guard of the United States. Our respective merits and defects provided a curious and almost exact parallel. With our rusty, if venerable, eighteenth-century systems, we both took between two and three years to create our national armies, and during that period an enemy could have done with us pretty much as he liked had we been alone. We too often forget the fact that French, Russians, and Italians stood in the breach while we English-speaking people were improvising our armies.

After the war America went ahead, at least in her conception of how national defence should be approached. We scrapped all our war machinery. America attempted to preserve hers. We restricted ourselves to our old eighteenth-century type of army, and, in fact, produced an article far inferior to that of 1914, not to speak of all the muddling and messing and disbanding which followed the war, while America tried, and claims successfully to have tried, to preserve her war-time armour, or at least to keep it bright for the day of trial. The practical result is that we shall take nearly as long as before to create a national army, while the American General Staff say that they can be ready in six months.

Like ourselves, America has had two lessons, but, unlike us, she only neglected the first lesson, and profited by the second. All her lessons of the Civil War were neglected, but not so those of 1917-18. She went back, as we did, to the small regular army in peace time, but prepared to reinforce it in time of emergency by such additional citizen forces as the emergency might dictate, and clearly foresaw that such forces would be based upon compulsion. The American National Defence Act of 1920 provided an excess of establishments, both officers and enlisted men, to be employed upon the organisation, administration, and development of the two lines behind the regular army, namely, the National Guard and the Organised Reserves, and made further provision for the Reserve Officers Training Corps, and the Citizens Training Corps. 'It was the President's desire,' said Mr. Weeks, the American Secretary of War, 'that the regular army should be so organised as to carry out this intent of Congress to the fullest extent.'

New Military System

The real interest for us of the new American system resides, of course, in the development of the second and third lines. The plan was worked out by General Staff Committees acting in conjunction with National Guard and Reserve Officers, and the basic plan was to reconstruct, so far as practicable, the fighting divisions of the World War, whose numbers, units, spirit, and traditions America desired to revive and continue. The design is of much larger scope than most people in England appreciate. The Regular Army supplies 9 divisions, the National Guard 18, and the Organised Reserves 27 divisions of infantry. Thus the first two categories supply three field armies of 27 divisions in all, and the last category another 27 divisions, making a grand total of 54, besides corps and army troops, including cavalry and aircraft.

The attached map shows the army and army corps areas of the United States Army and the positions where the National Guard and Organised Reserves divisions are in course of completion. If and when this work is completed there will be no need for reorganising on mobilisation, which unhappy process so marred and delayed the work of creating an army in 1917. All the old National Guard divisions of the war-time have been reconstituted, and three more have been added to them. These National Guard divisions are numbered from 26 to 45 inclusive. The individual States have much to say to the creation of these larger units, and, in particular, the provision of officers rests mainly with them. But the States and the War Department now co-operate closely. The latter provides the States with 900 Regular officers to keep the National Guard divisions up to the mark, and there is the definite policy, when the force is drafted, of federally recognising the National Guard officers. This policy is indeed already in operation, and several thousand of such officers have already secured such recognition.

The American War Department, having assured this first part of the new organisation, turned its attention to the Organised Reserves, which will take some time to put in order. Differing from the National Guard, the Organised Reserves are a purely Federal force, entirely in the hands of the central Government under powers granted to Congress by the Constitution (Article I., Section 8). The members of the force have a war obligation only, but in peace may be called out for fifteen days or for longer if they wish it. Special Regulation No. 46 of 1921 deals fully with the organisation, administration, training, and mobilisation of this force, while Regulation No. 48 covers all the same ground for the Enlisted Reserve Corps, and No. 44 for the Officers Reserve Corps, both of which enter into the composition of the Organised Reserves.

As soon as all these regulations were made—and they were not made hastily—the corps area commanders sent out officers to perfect the plans, and the writer found and observed this process in operation during his visit to America. It recalled the initiation of Lord Haldane's plan for our Territorial Force. Some 1,200 Regular Army officers are detailed for this work, and unless Congress starves the appropriations there is no reason why this great work should not soon be well on the road to success. It will be understood that these Organised Reserves, destined to form the second 27 divisions of the National Army, are nothing but a skeleton army in peace, cost very little money, and count upon compulsion to fill their ranks in an emergency. But all will be prepared for mobilisation, and when the units are filled up they will at once begin their training under trained officers. It will be a localised territorial force, in which each officer and drafted man will have his place assigned, and in time we must suppose, the allotment of clothing, equipment, and armament will answer to the rest of the organisation. It should be the same for the drafting arrangements. Each State will furnish its own troops in which it will take pride.

Training Reserve Officers

The Reserve Officers Training Corps is doing wonderfully well, and 90,000 students completed the year's training in 1921. The Citizen's Military Training Camps also proved so popular that little more than a quarter of the applicants could be accommodated. Industrial mobilisation has been taken in hand by the Supply branches under the clauses of the Defence Act which deal with it. These are very thorough. When war is imminent, the President has full powers to make contracts which then become obligatory and have precedence over all others. In matters of arms, chemicals, supplies, and so forth, the President is armed with the powers of a Dictator. The Defence Act allocated $20,000,000 'out of any money in the Treasury not otherwise appropriated' for the object of enabling the President to carry out these purposes. What an explosion there would be at Westminster were such a sensible arrangement embodied in our mobilisation plans!

Such is the big scheme. A Regular Army for the same general purposes as ours, but with a superabundant provision of professional officers, experts, and specialists ready to devote themselves to the constitution of reserve formations. Secondly, the National Guard divisions, which have military traditions and promise to take the field early and to form two-thirds of the first three armies after amalgamation with the Regular Army divisions. Thirdly, the Organised Reserves, a skeleton army in peace, which will take time to perfect, but when perfected will double the field army of the United States. A total of 54 divisions, with Army and Corps troops on a modern scale, and the whole representing not the limit of what America can do, but what she is prepared to do in the first instance if a national emergency is declared by Congress to have arisen.

It needs only to be added that an investigator who studies the National Defence Act of 1920 will find a difficulty in understanding the system from this Act alone, because it is drafted in the form of amendments to the Act of 1916, and this difficulty is enhanced by the fact that details of organisation are not specified, being left, as they very wisely are, to the discretion of the President. For the time it is only at Washington, and by the courtesy of the United States General Staff, that the scope of the new basic plan can be understood. But it is clear, after such inquiry, that Colonel John McA. Palmer was perfectly justified in saying, as he did in a recent lecture at the Army War College, that 'it is primarily the object of our new law to perpetuate the framework of the organisation developed in the World War, so that its tremendous cost can be funded as a permanent investment for all time.'

Observations

It has been shown that the United States contemplates, in case of emergency, the constitution of six field armies of 54 divisions of infantry in all, and assumes that the selective draft system—or, in plain words, compulsion—will fill the ranks when the necessity arises.

It is not my intention to criticise the manner in which the basic plan of the American General Staff has been influenced by the action of Congress in regard to appropriations. Congress, like Parliament with us, holds the purse strings, and we know only too well that when politics demand retrenchment of expenditure the army is always the first to suffer. At such moments, promises, commitments, responsibilities, and regard for the future weigh not a feather in the balance. But there is this to be remembered in the case of America that even when meagre appropriations appear to render futile the hope of carrying through a military plan there is always in reserve the lofty patriotism of the American citizen and the local patriotism of the individual State. These are serious influences, which counteract parsimony in Congress, and are, in fact, in process of doing so.

The important thing for us is to ask ourselves whether the principles of the American plan are applicable to our conditions, and if so whether we should not set out on a new course and endeavour to apply the American principles to our own case. There is no great Power in the world less open to attack than the United States, yet after a full review of all the conditions of the world and two years after the World War had ended, the Americans deliberately decided not to be caught napping again, and determined, not indeed to raise forthwith the great armies required on an emergency, but to prepare by voluntary effort the framework of a great military organisation, with the least possible demand whether for service or for money, so that the frightful waste and delay of 1917 should not again render the country ridiculous, impotent, and almost disarmed in face of foreign pretensions and aggression.

The Army and Foreign Policy

There can be no doubt that the American plan, when completed, will add immense strength to American policy. No longer will there be foreign General Staffs to assume with levity that a war can be concluded before America can take the field, or to hug themselves with the idea that as there is no 'militarismus' in America, so can there be no military force. The years 1917-18 exploded that illusion, and the new plan reminds those whom it may concern that the intervention of America in disputes which affect her interests can be more swift, fierce, and sharp than when in 1917 she was learning to muddle through. There is also in the plan an implied understanding that compulsion will again be resorted to, and such thought conjures up the memory of the four millions of Americans in uniform and under arms in 1918, and of the ten millions or more behind who were awaiting the order to join. At the same time the plan is no threat to anyone. Compulsion in peace time has been set aside, and all that has been done has been to organise mobilisation so that the disorders and vexations of unprepared war may no longer hamper and enfeeble American diplomacy.

After all, no State in Europe has disarmed, except under compulsion, but England. The States that have disarmed under compulsion, namely, Germany, Austria, Hungary, and Bulgaria, desire to revert to compulsory service and will do so at the first opportunity. We hear constantly of new military laws even in little countries like Holland and Denmark. As for the rest, they have larger armies than ever, and most of them are much better equipped. America's voice will be heard when she chooses to speak, and the existence of this great military plan will add weight to her words. She has found the happy mean between armament and disarmament. A country like England, which disarms in an armed world and still aspires to make her decisions respected, is asking too much of fickle fortune. When acts follow words, might talks. Words then count no more than the cooing of pigeons in the woods.

Comprehensive and Far-seeing

If we look into this American plan we must certainly congratulate General Pershing, Major-General Harbord, and the very capable General Staff under them for having devised a plan at once so comprehensive, so far-seeing, and so suitable to the genius and circumstances of their people. There are few more

Regular Troops under arms than before the war, but there are now Regular cadres allocated to the building up of the second and third lines, and there is good provision of Reserve officers for these lines and of a training staff. The distribution of the corps areas and of the divisional headquarters shows that full use will be made of all the resources of the territory, and the railway system has not been left out of consideration.

The writer has looked over the establishments of the different categories of troops, and is impressed by the care and industry displayed in their compilation. They are too detailed to be recorded here, but it may be said generally that they appear to be very complete. In addition to the larger formations already alluded to there are 12 divisions of cavalry in process of construction, for which there will be some 54 cavalry regiments eventually available. The Organised Reserves, in addition to the three field armies found by them, constitute a kind of pool whence any deficiencies in the other lines are made good.

From the same reserves there are drawn 139 motor transport companies for the corps trains and 59 motor repair sections, while many elements of the Air Service are also found by them. This Air Service includes six headquarters and headquarters squadrons, 18 communication sections, and 18 airdrome companies, 6 wing headquarters, 12 pursuit groups, 6 attack groups, 24 observation squadrons, 6 air parks, and 12 photographic sections. Approximately one-half of the Air Force will be found by the Organised Reserves. The Corps artillery consists mainly of 155mm. guns and howitzers. The Tanks form 36 battalions, of which 30 are to be found by the Organised Reserves. It will naturally be recalled that the American Army found itself in possession of an abundant supply of war material of every description at the close of the war.

The President and his War Secretary are presiding over the creation of a great national institution—that is to say, one which will expand a small professional peace establishment into a great and largely non-professional war army. The sacrifices of the war generation are being made a fruitful heritage for the future. American officers hereafter will be familiar with the large organisations and problems of war, and will disseminate their knowledge far and wide. The necessary connection between the army and the business world will be maintained, with the broad result that months if not years of time and oceans of money will be saved in the preparatory period before the army takes the field. These are great results, and results worth attaining. That they can be attained is certain if Congress remains firm and the President supports his War Secretary as well as he has hitherto done.

1917-1923

On January 24, 1923, that fine soldier and good comrade General Allen hauled down the Stars and Stripes from Ehrenbreitstein and left France the poorer for the loss of a good friend. It was the end of a great adventure.

As we look back over those eventful years we remember how the United States gradually became convinced that her place was alongside the Western Allies and how the old love of isolation gradually broke down under stress of many circumstances, not least of which was the incorrigibly stupid diplomacy of Germany and her apparently complete indifference to American opinion. There are some who still think that the sinking of the *Lusitania* and the declaration of ruthless submarine warfare were the only two cases upon which the United States was justified in declaring war, but there was all the time continuous aggravation, so that when the break came it came with the enthusiastic approval of the whole country, which had become embittered and excited by a whole series of hostile acts which the people could no longer put up with.

Few people in Europe expected much from American military support at first. The United States was rather regarded as a

Treasury and an immense storehouse, invaluable to replenish the half-exhausted supplies of the Allies. Her potential military power was scarcely understood, and at the best it was thought she would have to pass through all our long and bitter experiences before she could place a respectable army in the field, for nothing had been done to prepare for such a deployment of force before the break came. Yes, this was true. America lived painful days and made as many mistakes as we did. The need for haste was not then understood by us, nor fully impressed upon America, until after the smash of March, 1918.

Pershing reached Paris on June 13, 1917, and for a time he and his staff were completely overwhelmed by the magnitude of their task and by the recognition of war conditions with which their army was totally unacquainted. Yet three weeks later Pershing cabled to Washington that 'plans should contemplate sending over at least a million men by next May,' and the American Administration began to work on those lines, while Pershing's staff began to prepare for the arrival of these numbers. That year 1917 was not a good year for us on land or sea. The great French offensive failed, Italy was heavily stricken, Russia broke down, and in the five months ending June 30 three and a quarter million tons of Allied shipping were destroyed. It was not an extravagant hope for the Germans to entertain that they could transfer their eastern armies to the west in 1918 and end the war by victory. We must recall those days if we wish to remember the value and timeliness of American help.

The strenuous work accomplished in the course of the succeeding twelvemonth by the American Administration at home, and by the American Army and Navy in Europe, must be regarded as an effort almost unparalleled in the annals of war. Organising, fighting, and training all went on simultaneously, but owing to lack of shipping the required numbers could not at first be supplied. So when our defeat occurred in March, 1918, there were only eight American divisions in France, of which only two were in the line.

Only by efforts of an unprecedented kind were arrivals hastened. By June 1 there were 18 divisions in France, by August 1 32 divisions, by November 42 divisions, and as these divisions were very strong there were 1,718,000 American troops in France, including non-combatants, and at the armistice Pershing was preparing to keep up 2,000,000 men and making plans for the arrival of 2,000,000 more. Had the war been prolonged into 1919 the American Army would have been the preponderant force on the Allied side.

This splendid military effort, the great activity of the American Navy, and the invaluable support of Americans in almost every department of State affairs were, however, only a part of the services rendered. The real splendour of American comradeship was best seen in the cheerful, uncomplaining, gallant way in which the officers and men of the United States worked in with their Allies and subordinated all their national aspirations to helpfulness towards their hard-pressed friends. They were real good comrades in the war time, men whom we shall for ever remember with affection and esteem, while as for their splendid rank and file the crusading spirit was upon them and they were hard to stop. They had their troubles like the rest of us, but they all played the game. They thought that they won the war, but so, again, did all the rest of us, and history will probably assign to no army the most conspicuous share in a victory gained by the common effort of all.

Well, the flag is hauled down at last, and they are gone. It is more wonderful that they should have stayed so long. As one travels through the great Middle West Europe seems very far away, and one takes daily less and less interest in European affairs. We are in a huge continent with unlimited local interests, and with openings on all sides to attract energy and enterprise. The divisions, the troubles, and the cries of Europe reach us in faint echoes, and do not seem to concern us very much. Europe appears to be a very cantankerous continent, and to consist of a pack of little States unable to agree. So the American turns to his own affairs, and only a minority understand how much the Western farmer and the Eastern manufacturer depend upon their European customers. The United States cannot quite decide whether they have a foreign policy, and what that policy ought to be if they have one. The constitutional system which gives no Minister a seat in Congress, and gets along without a Government bench, is antagonistic to the American comprehension of foreign politics and to the pursuit of any clear and consistent policy.

So it was that the Versailles Treaty, which America so largely fashioned, the League of Nations, which was an American child, and the Anglo-American guarantee, which meant life to France, all went by the board, and the most splendid gesture of American history was robbed of its reward by the post-war attitude of America. Yet so long as the Americans remained at Coblence there was an outward and visible sign that America, or some people in America at least, were thinking of the war fought and of the peace signed in common, and, so long as they remained, all hope of American co-operation in the hard work of peace was not abandoned.

It is a tragedy that they are gone. Yet somehow their soul remains and 'still goes marching on.' No matter what happens, the remembrance of this great American adventure, the recollection of the whole-hearted co-operation of a great people in a war fought for right and justice, and the memory of all those fine fellows who made light of difficulties as they did of dangers, will never be effaced from our minds.

CHAPTER VI

The French Theory of War Preparation

The *Conseil Supérieur de la Défense Nationale* the Supreme Authority—The President's Powers—The 1913 Conception—The Decree of July 1, 1921—The Decree of November 17, 1921—Immense Task Thrown Upon the National Defence Council—The Seven Ministers Responsible—The Commission of Studies—The General Secretariat—The Machinery for Organising National War—To be Completed by 1924 or 1926—A Generation May Elapse Before the Nation Habituated to the New Plan—Difficulties—The Scheme Somewhat Cumbrous—Who Initiates Anything ?—Secrecy Not to be Expected—The Plan of the War—The Last Word in National Defence—Decentralisation of Control the Strong Point.

Now that we are by way of overhauling our higher councils of defence, it is well that we should glance at what our French friends have been doing lately in the same order of ideas. This is necessary for two reasons. The French are sharper fellows at theory than we are, and it has always been true that when we can combine French theory with British practice we get the best results possible. Also, if by some unfortunate chance we are plunged into some fresh trouble alongside the French, it would be an undoubted advantage that the two Governments should have similar ideas with respect to national mobilisation and the conduct of a great war, for then co-operation would be much easier and also more rapid.

The *Conseil Supérieur de la Défense Nationale* is the supreme authority on defence questions in France, though it does not, so far as one can judge, deprive the President of the Republic of the wide, if slightly vague, powers conferred upon him under Article

3 of the *Loi Constitutionelle* of Feb. 25, 1875. This lays down that the President '*dispose de la force armée*,' but adds that he may not declare war without the previous assent of the two Chambers.

There is a *Conseil Supérieur* for the Army and for the Navy. Both are important, but the Council of National Defence is now predominant, and has the duty of examining all questions which require the co-operation of several Ministerial Departments It is not a new creation. It was born by a decree of June 14, 1913, but little was heard of it during the war or immediately afterwards, and in fact an organisation of this kind can only reach its full development after years of careful study and continual labour. Briefly, it was, in the 1913 conception, a council of six Ministers qualified to give advice inter-ministerially on defence questions, and it was laid down that all such questions were to be studied and prepared by a *commission d'études*, doubled by a general secretariat. Actually very little was done.

Soon after the war the 1913 idea was taken up again. The first change was the addition, by a decree of July 1, 1921, of an aeronautical section to the council, with somewhat wide powers and a secretariat of its own. This began to introduce an inconvenient complexity into the arrangements, and the definite place of aeronautics in French war councils was not then, and possibly is not now, finally determined. Our solution of an Air Minister with a seat in the Cabinet appears a good plan to ensure co-ordination, and his place would be marked in the National Defence Council of the French, or in any similar body of ours, whether in peace or war. But France has not yet a Minister for Air.

As the 1913 decree began to be studied once more, various improvements suggested themselves. The National Defence Council was much altered by a new decree of Nov. 17 in 1921. There was imposed upon it the duty to say how the governmental and administrative mechanism should be conceived and organised in time of war; how the energy of the country should be distributed between military efforts and organs dealing with the life of the nation; what part of the agricultural activity, the commercial liberty, and private production should be maintained; what means of communication, of exchange, and liaison should be left at the disposal of the public; what *rôle* should be given to the intellectual and professional formation of youth; and, in short, by what means the country should conciliate the necessity for fighting victoriously with the obligation of preserving its future existence. In order to enable the council to carry out this work the Minister of Public Works was added to the six Ministerial members of the 1913 plan, making seven in all, while the vice-presidents of the Army and Navy Councils were added, with consultative voices only. The council remained Ministerial, and its members now are the Prime Minister and the heads of the Ministries of Foreign Affairs, Interior, Finance, War, Navy and Colonies, and Public Works.

The *commission d'études* was also greatly enlarged and its duties defined. It now includes some twenty-one or more representatives of all public departments, and its duties are to study all questions which have to be submitted to the deliberations of the National Defence Council. The commission works under the high direction of the Prime Minister, who nominates the vice-president. The Prime Minister divides up the *commission d'études* into four sections, which deal with:

1. The general organisation of the people in time of war.
2. The conduct of the war.
3. The victualling of the country and transport of all categories.
4. Manufacture of all kinds.

The aeronautical section now forms part of the *commission d'études*. The Council of National Defence is authorised to hear anyone who can throw light upon any deliberations in which it is engaged. At the time of the above changes various well-known men were placed upon the *commission d'études*, such as Generals

Buat, Mangin, and Ragueneau ; Admirals Grasset and Marguerie ; and M. Emile Durand of the Sûreté. All those were ex-officio members, while the Ministries were represented by such men as M. Philippe Berthelot, M. Théodore Tissier, M. Denoix, MM. Mahieu, Fighiera, Fontaine, and others. The secretary-general of the National Defence Council was added later to the *commission d'études*, with consultative voice.

The composition of the General Secretariat appeared in the *Journal Officiel* of January 5, 1922. It now includes twenty-six officers and functionaries of the various public departments, and is divided into four sections, corresponding with those of the *commission d'études*. The secretaries of all the sections are officers of the Army or Navy. There is, in addition, a subaltern personnel of clerks, typists, etc. The secretary-general is General Serrigny, an officer of much ability, well known to those who visited Marshal Pétain's Grand Q.G. during the war.

Such is the new French system for the preparation of a national war. It may be admitted that, so far as human ingenuity can cater for the unknown, it provides the machinery for organising national war. The French council can turn itself into a war council at a few minutes' notice and go ahead with its war. But —General Serrigny has himself admitted as much—the study involved cannot be carried out to the point of perfection without a parellel knowledge of the mobilisation arrangements of allies, and of the intentions of enemies, actual or eventual. The plan must be flexible enough to respond to the fluctuations of inter-State relations, and General Serrigny was probably right in assuming that the programme would take from three to five years to work out, and that a generation might elapse before the habituation of people to plans for national mobilisation would be complete.

There are other difficulties. It is the rarest thing in the world for the conditions of any war to be foreseen, and arrangements for national war must meet the case that actually arises, and not a case that does not arise. How various are the conditions which both the French and ourselves may have to meet in future wars in Europe, Asia, Africa, or America ! A war may be continental or maritime, or both, and it is not given to man to prophesy where, or against whom, a war will take place, nor so to arrange national mobilisation that it will meet all cases. Our friends, like our enemies, may be in any state of preparedness or the reverse, and in these days, when common interests tend to make a war world-wide, who can say whose toes we may not tread upon and force into the field ? There is relative security only in preparing to meet the most dangerous contingency foreseen.

It must also be said that the French scheme is somewhat cumbrous. It is not clear who initiates anything. If anyone has a bright idea he has to get it to the general secretariat first, and when they have ironed it out it may get eventually to the *commission d'études*. Then a fresh examination begins, and, in the rare event when anything further happens, the bright idea comes before seven civilian Ministers, who may not understand it, or may find a score of political reasons for burking it. The war may easily be over before the bright idea is hatched out. As for secrecy, for which our neighbours, with all respect, are not famous, how can it be expected, when every paper is duplicated, triplicated, and quadruplicated on each stage in its course through the triple barrier of organised French bureaucracy ?

Seriously, will strategists pass all their plans through the second section of the *commission d'études*, which deals with 'the conduct of the war' ? We can feel sure that not one of the French marshals will do anything so silly, and it is also certain that General Serrigny will not try to make him do it. The plan of a war must be kept in the smallest possible number of hands. Nothing that conflicts with this need can live for long.

We have not organised national war at all, and the French are over-organising it. As one thinks of the fifty-four members of these three bodies who for the greater part have nothing to do but organise national war, one would, if a Frenchman, tremble. Awful will be the paperasserie that will accumulate, and most of it, if we have a war in the Pacific, will be useless. But one must admit that if this French scheme is the last word in national defence, it at least kills dead the theory of a Defence Minister. It is clear that he could have no place in such an organisation and would be an unmitigated nuisance. It is a strong point of the French plan that instead of centralising control it decentralises it and brings in all departments of State to take their part in a national mobilisation. With that aspect of the plan we must endeavour to conform.

But when General Serrigny asks the French Military Attachés in London to report upon our mobilisation plans, and to say what help we can give to an ally, or what help we need from him, so that the plans of the French Council of National Defence may become peerless, I feel quite sorry for the gallant officers concerned, for an England more destitute of defence than this England there never was. We do not mean to hustle any Prime Minister, but we look to him to retrieve our lost security. It is not all a question of money. The things which we want most, after an adequate Air Force, are the renewal of registration, and the redrafting of an improved Military Service Act, which, when replaced on the Statute Book, can be put aside till needed, like the old Militia Ballot Acts were. These will cost little money, and the whole thing could be put in order in a very short time. That would be an encouraging event, and would make friends and enemies understand that we were not going to let anyone fool us.

CHAPTER VII

The "Black" Troops of France

The Population Question Between France and Germany—Methodica Organisation of the French Colonial Empire—The Beginnings of the French Colonial Army—Participation of Native Troops in French Wars After 1836—845,000 Natives Engaged in France During the World War—The New Organisation—Native Regiments Quartered in France—The Mother Regiments in the Colonies—Social and Political Aspects of the Question—Military Merits of the Plan—Not Wholly a French Question.

THE peace of 1919 left France with a population of nearly 40,000,000 souls in her metropolitan territory, and Germany with a population of 60,000,000. But, while the German colonies had disappeared, the colonies of France had been increased, and they now form an oversea empire of nearly 60,000,000 souls, making 100,000,000 of people in all under the French flag. Given the solidarity of the Reich, it was a natural conclusion that, could France methodically organise her colonial subjects for war, she need no longer dread the future. This purpose she is in process of carrying out, and the measures which she is taking to this end are important and significant, even though France has no self-governing dominions, and consequently her situation and ours are not identical.

The French colonial army was born in North Africa in 1836, and native troops from that part of the world took part in the campaigns of the Crimea, Italy, and Mexico. They also fought in 1870-71, and those who have carefully examined the accounts of the first battles of that campaign must pay a tribute of sincere admiration to the bravery and dash of the native troops engaged. In those old days, and even up to the outbreak of the World War, the native troops of colonial France mainly fought their own corners overseas, and their soldierly conduct on many a hard-fought field was conspicuous. So it came about that they were engaged again in France in 1914-18.

At first the numbers were small, and Marshal Joffre told me in the spring of 1916 that there were then only 100,000 native troops in France. But the French and Allied penury of men later caused a greater effort to be made, and at the close of the war no less than 845,000 natives had done service in France, of whom 535,000 were soldiers and 310,000 labour contingents. The bulk of the native soldiers came from North Africa. There were 181,000 Senegalese, 175,000 Algerians, 50,000 Tunisians, and 34,000 from Morocco. Indo-China and Madagascar also sent large contingents. The losses were considerable, and of North Africans alone the killed numbered 28,200.

It has been decided to organise seriously the French colonial domains for war, and to include in the peace garrison of France no less than twenty-four native infantry regiments as the spear-head of others to follow if the sea is clear. The present organisation of the colonial forces includes 10 white regiments of infantry, 30 native regiments with white cadres in North Africa, and 39 in other colonies. In peace time 7 of the white colonial regiments and 24 native regiments with white cadres will be permanently located in France. The total home and colonial peace establishment of the French Army is 660,000. That of the colonial troops is 189,000; of whom 91,000 are of North African origin and 98,000 in other colonies. That is to say, that native troops now number nearly one-third of the French peace establishment.

In most colonies conscription is the rule, but Morocco is for the moment an exception. It is anticipated that, without any great effort, given the three years' colour service and seven years in a reserve, which are the usual rule, there will be in ten years' time 400,000 trained native troops available and 450,000 ready to be trained in those 'mother regiments' which France proposes to keep in her different colonies. The encadrement of these large forces is a big affair, as the usual French practice is to have not only white officers but many white N.C.O.s in the native regiments, but the French believe themselves to be capable of supplying all the necessary specialised white cadres acquainted with native races and idiosyncrasies.

The most important novelty in these arrangements is the permanent quartering of 24 native regiments in France during peace. This is the limit of what is considered advisable and not of what is possible. With the present eighteen months' service of Frenchmen, and an annual contingent of 250,000, it is not practicable to have more than 32 divisions formed in peace. The native troops, with the white colonials, find six of these divisions. They are quartered mainly in the north-east, in the occupied territory, but Lyons, Marseilles, and Bordeaux will also know them. It should be said that the 32 divisions of the present visible French peace establishment are only a frontier or covering army, behind which the armies on mobilisation will form in all security. What number of divisions will be created in war time is the secret of the higher French command, but it will certainly be much larger than 105, which presented the number in the field at the close of the war. The divisional unit still remains the usual standard of comparison, but in France the infantry itself is only 50 per cent. of the peace establishment, and the number of rifles in a division is less than it was. Very largely, the French Army has become an army of *matériel.*

The native infantry units in France have a high peace establishment to enable them to take the field at once. They leave behind them a depôt in France, where their reservists will assemble to replace casualties in time of war, and it is naturally the case that the arrival in France in time of war of the largest numbers that can be spared from the colonies is being carefully worked out.

There is the military aspect of this question to be considered, and also the social and political aspect. On the military side, so far as numbers are concerned, the object which France set out to achieve may be secured. No doubt native troops have their weaknesses. To bring them to Europe and engage them during the winter on the defensive in muddy or frozen trenches, as we engaged our Indian Army divisions in 1914, against a highly efficient European enemy, better provided with heavy guns and other material for trench warfare, is to ask for trouble, and is only permissible when we can do nothing better. But, as we know with out Sudanese, there are native races which provide admirable missile troops in attack, and when they are properly used by people who understand them they are first-class military value. A system which may within ten years provide a million native soldiers for France, to make up for the difference between the French and German home populations, cannot be ruled out by colour prejudice.

So far as I can ascertain, and although the character of some of the North African races makes the individual a *mauvais coucheur* at times, the French people accept their native troops, and some towns, failing a white garrison, ask for a native regiment in their midst. The presence of a garrison in a town brings so much grist to the mill that the desire is comprehensible, but I am fairly confident what our people would say if we proposed to quarter permanently 24 native regiments in Britain. In theory the French Republicans propose to accept the native as a man and a brother, and, as General Degoutte on the Rhine has enforced this principle upon me with serious arguments, I cannot doubt the honesty of the opinion. In practice, however, my observations have shown no fraternisation between French and native troops in any garrison, while among other European people the principle is not liked, and, of course, by the Germans it is loathed.

On the social and political side only the future can disclose the consequences of these measures. Discipline will be maintained provided that there are no collisions or ill-feeling between the white and coloured people, and provided also that the natives are not discontented with the length of their colour service, want of furloughs, and inadequate pay and separation allowances. It is possible that the natives may contract habits of drinking, of which complaints have already been heard, but the greatest danger is the loss of the prestige of a dominant white race on account of relations between the natives and the lowest class of white women and other sorry individuals. To regard the native as having all the rights of a Frenchman is beautiful theory, but, if he has such rights, he may inquire why his colour service should be twice as long as that of his French fellow-citizen. The French colonial population is loyal, but the people, with some exceptions, are conquered aliens, and may take home, and spread abroad, not only the vices of the lowest order of French people, but the seed of subversive political ideas which will lead to much trouble hereafter in the continent of Africa.

One must acknowledge the military merit of the new scheme, but the permanent quartering of large bodies of natives upon a white population appears to contain more future dangers than it offers present advantages, and when a settlement with Germany is made France would be well advised, in her own interest, to keep her native troops in their colonies of origin and not in France. Is it really worth the social and political risks involved to keep this mass of native troops in France when they could be carried across from Africa so speedily at the first threat of war?

This is not wholly a French question, as it seems to be on the surface. Any disturbance arising in North Africa owing to political upheavals or rebellions might quickly spread through the African continent and affect the interests of many people besides the French. It is mainly on this account that we have to watch this new and interesting experiment with careful attention.

CHAPTER VIII

French Aerial Policy

Reasons Why the French Have No Air Ministry—Aviation Has No Theatre of Operations of its Own—As a General Rule it Acts With and For the Army and the Navy—Generals Can Do Nothing Without Aviation of Reconnaissance and Observation—Fighting Squadrons Have to Protect Observers and Disturb Those of the Enemy—Bombing Squadrons Complete the Action of Guns—An Army Commander Must Command His Air Forces Which Are an Arm Like Any Other—Independent Air Forces Must be Under the Commander Responsible for the Conduct of the War—French Army Reductions Dependent on a Large Provision of Aircraft—The French Regulations—Air Squadrons with Divisions, Army Corps and Armies—The Aerial Divisions—Organically Part of the Army Reserves—France Has the Air Force of Her Army and Her Navy—We Must Have the Same—We Do Not Possess it Yet.

It is necessary that we should keep in mind the reasons why the French have not yet accepted the principle of an Air Ministry, and the arguments which are used to support this decision.

These arguments are given in the May, 1923, number of the *Revue Militaire Française,* which is published with the support of the French General Staff, and we shall not be very far wrong if we suppose that the arguments used represent those of the most authoritative military circles in France.

Lieut.-Colonel Pagecy admits that there is almost unanimity in the French Press in favour of an Air Ministry, and that it is an ungrateful task to advocate a contrary view. But he says that, as this contrary view is believed, rightly or wrongly, to be most conformable with the general good, he will endeavour to defend it.

Aviation, he says, has no theatre of operations of its own. It may have to carry out distant bombardments with a certain independence, but as a general rule it acts for the profit of the Army and with the help of the Army, or for the profit of the Navy with the help of the Navy. Aviation must have its own organisation and statutes, but it is an arm amongst others. Aviation of reconnaissance and observation plays the greatest *rôle* in war. It is the eye of the commander and of the artillery. Our generals in all commands can do nothing without having this type of aviation under their command, a type which must know thoroughly the tactics of other arms, and must be able to see clearly amidst the disorder of a battlefield. It is also the eye of the artillery, which is vowed to impotence if its fire is not directed and regulated by an aviation acting in close touch with it and acquainted with all its methods.

The fighting squadrons have for their chief mission to protect our observation and disturb that of the enemy. They cannot do either unless they are placed under the orders of the Air Commander and of the Commander of the Army. The bombing squadrons complete the action of the guns. The mission of the two is so closely connected that some Germans wish the bombers to be under the orders of the artillery commander.

Many officers, says Colonel Pagecy, look upon the *aviation de bataille* as troops of assault, launched against the enemy in successive waves and playing its part in a battle like an instrument in an orchestra. How, then, can we not place it under the conductor of the orchestra ? In old wars the Army Commander had his infantry, cavalry, and artillery. Now he has his infantry and tanks, his artillery, and his air forces. Therefore the Army Commander must command his air forces. It is not enough that they should operate in liaison with him ; they must be under his orders. Being under his orders in war, they must also be under him in peace.

As for the independent air forces which execute distant

bombardments, they also must be under the commander who is responsible for the issue of the war and is alone in a position to decide upon and to choose his objectives. In short, an army without an air force is no longer an army, nor is a navy without its air force a navy, and it is as improper to remove the air force from an army and a navy as to take away their cannon and their rifles.

Such is the case of the French Army against an Air Ministry.

We must glance at the French Army organisation in order to understand the whole position.

The French ended the war with something over 100 divisions. By successive reductions this number has now been reduced to 32, and the reduction has been possible solely because a generous provision of aircraft, guns, and tanks has enabled our neighbours to substitute machinery for man-power. People were naturally alarmed when they read of 140 French air squadrons in the Duke of Sutherland's speech on March 21, 1923, and pictured these squadrons as an independent air force, capable of going anywhere and doing anything to make our flesh creep. Actually—and someone in authority should have stated the fact—the French have an army policy and the aircraft of this army. As we have neither an army policy nor aircraft we find it a trifle hard to understand this matter ; but it is all the same true that we can take as little objection to the French proceedings as they could take to ours had we a navy policy and the aircraft of this navy.

Let anyone study the French 1922 Regulations on *L'Emploi Tactique des Grandes Unités* and the real situation will be comprehended. There it will be seen that each French division of infantry includes in principle one observation squadron of aircraft and one balloon company. Each French army corps, it will also be observed, has one or two similar observation squadrons, and each army has others. The air squadrons form as much part of the division, corps, and army as do the other services of each formation. There are, in addition, the important aerial divisions, which each include a bombing brigade and the necessary fighting squadrons to protect the bombers during their missions. They are, of course, formidable instruments, which can be employed anywhere and for any purposes, provided that they have the necessary aerodromes at disposal; but these aerial divisions, again, in principle form organically part of the general reserves in the hands of the commander-in-chief, who places them at the disposal of his armies in war according to the requirements of the moment and his plan of operations.

I have an impression that after detailing 32 air squadrons for the divisions, another 32 for the army corps, and others for the armies and for G.Q.G., the French cannot have, for the moment a great number of aerial divisions to show. These can be created later, when the existing 1,260 aeroplanes rise to the expected 2,160 of 1925. That would, of course, be a big affair, and it is explicable by the fact that the number of French divisions after a general mobilisation will be very largely increased and that provision must be made for them. It is in any case fair that we should admit that the present French air force is organically part of the field armies, and cannot be diverted to other uses without materially weakening those armies as at present constituted.

I consider that this matter should be more fully explained to Parliament than it yet has been. Bald statements of total numbers do not enable the public to form a judgment. We want not only the numbers but the character and cost of all these planes and their intended distribution. To suppose that a full statement cannot be given is to rate our General Staff and military attachés as incompetent. The main expenditure is distributed over the French army, navy, and public works estimates, but there is no secrecy about these things, while we and our French friends have few secrets from each other. I am personally dubious whether the French air force costs anything like what our equivalent of it is reckoned to cost, namely, £35,000,000, but

I imagine that our air representatives will be closely examined on this point later, when all the facts should come out.

We should accept the fact that France has the air force of her Army, and that we can neither challenge nor object to it. Many people grumble at and criticise the strength of the French Army. When all the lions and all the lambs lie down together things may change, but without the French Army of to-day Europe would be a seething cauldron of warring communities. A nation which has been so predominant in a military sense on the continent of Europe as France is to-day, thanks to her army and her alliances, and has exploited this predominance so little for its own selfish ends, I do not recall.

This same freedom of France to create the air force of her Army we must claim for ourselves in relation to our Navy French problems and ours are not the same. France looks to the land, and, looking there, accepts with a very wry face the Washington Treaty, which does not indeed alter anything of the French naval position, but does, in relation to capital ships, accentuate the fact and proclaim to the world that France is no longer a first-class naval Power. To those who know the glorious history of the old Royal Navy of France, and realise the French pride of race, it is needless to say how bitter is the pill. But as France has swallowed it, then must we at least admit as a compensation her undoubted right to the army of her policy and to the air force of this army.

We are not self-contained, like France, and we could not exist were our communications closed with the outside world. So we need the Navy of our situation and the Air Force of this Navy. We do not possess it yet. We have given away a great many things which we possessed at the close of the war, and among them the luxury of saying what we please about all foreign Powers. That luxury we have enjoyed for many centuries under the shield of a predominant Navy, and we have become so accustomed to its use that we do not see how circumstances have changed Our Navy is not predominant any more, and if it were it could not defend us in the air. The old style of laying down the law to foreign Powers and not minding their hostility if it came were better laid aside for a time.

The duty of our Navy is to protect our ocean trade routes and guard our shores against attack. It cannot do so now without the air arm of its needs. The command of the Channel has been considered indispensable for the political independence of Britain from very ancient times. Whether even this is now safe against attack by submarines, mining craft, and aircraft in co-operation is a matter for the Navy to say. That which is needed for home defence must be accorded without grudging. But there is a right way and a wrong way of stating and solving military problems. This particular problem should be investigated and announced as part of our naval policy, and not in any other manner. Let us cease to gird at France for having the air force of her army. Our object should be complementary, namely, to secure the Air Force of our Navy.

CHAPTER IX

Charles de Freycinet and Walther Rathenau

De Freycinet's Work for France—*La Guerre en Province*—Work under Gambetta in 1870-71—The Creation of Armies—A Great Improvisation —His Labours After the War—Points out French Weaknesses and Reorganises the Army Between 1888 and 1893—His Convincing Oratory and Reasonableness—Great Knowledge and Rare Assiduity —Rathenau—A Man of Devouring Activity—Encyclopædic Knowledge of Affairs—Services to Germany During the War—Reconstruction After the War—Had Not the Qualities for the Foreign Ministry —His Share in the Russo-German Treaty of Rapallo—The Genoa Conference—Bad Impression Created—An unsatisfactory Explanation—Rathenau's Writings—Quotations From His Most Popular Work.

M. De Freycinet

When a statesman dies at the age of ninety-five it is almost inevitable that he should be almost unknown to the generation of the moment, which is usually and singularly ill-informed about things that have happened during the fifty preceding years. To that rule M. de Freycinet was no exception.

M. de Freycinet's life and example both in the matter of organising war from untrained elements, and of reorganising the French Army afterwards, provide lessons applicable to the future, while his manner of handling a modern Parliament has rarely been excelled.

That short but masterly little book of his, *La Guerre en Province*, in which de Freycinet described the resistance of the French people to the overwhelmingly superior German armies in 1870-71, when the French armies had been beaten and invested, and no hope appeared to remain of victory, is certainly one that every soldier should possess.

When Gambetta took office on October 10, 1870, Paris was invested and Bazaine shut up in Metz. There remained, in the whole of France, less than 40,000 regular French troops, an equal number of national gardes mobiles, 5,000 cavalry, and 100 guns, the whole in a bad state and much distressed. Four months later, Gambetta by his fiery eloquence, de Freycinet by his organising talent, and Colonel Thouman by his competence in artillery matters, had created armies 600,000 strong, with 1,400 guns, the average production having been 5,000 infantry and two batteries of artillery a day. These forces, united with the 80,000 regulars and gardes mobiles above-mentioned, went to form the 12 army corps, numbers 15 to 26, the army of the Vosges, and other important groupings, which challenged the perfectly organised, disciplined, and equipped German armies, already 800,000 strong, with 2,000 guns.

By these splendid endeavours France regained the esteem of the world, and, better still, consoled herself in defeat by the thought that she had done her duty, and had not despaired in adversity. To that effort de Freycinet, the frail little figure, 'the little white mouse,' contributed much by his talent and vigour, and more by his extreme tenacity. But when the war was done he neither exalted himself nor blamed others. He sought out and exposed the causes of defeat, which he showed to be the numerical weakness of the French armies, the deplorable higher direction of the war, the inferiority of armament and equipment, bad organisation, indiscipline, insufficiency of staffs, supply services, and cadres, and the misdirection of peace strategy. As Gambetta said at Bordeaux on June 26, 1871, 'We have been beaten by adversaries who had on their side the advantages of foresight, discipline, and science, which things prove, in last analysis, that even in conflicts of material force it is intelligence which remains master.'

To the best of my recollection, the average duration of a French Ministry, from 1871 to 1914, was about ten months, and until M. de Freycinet went to the Rue St. Dominique as War Minister in April, 1888, no civilian had held this post since the year 1848. Fortunately for France and her army, he retained this post when Tirard's Ministry succeeded that of Floquet, kept it in his own Ministry, which came next, and still remained War Minister in the Ministries of Loubet and of Ribot which followed. Thus from 1888 to 1893 he remained in charge of the army, and even afterwards he was President of the Army Commission, and played a large part in the elaboration of the Two Years' Service Law. It is, perhaps, forgotten that he held office in the Briand Cabinet for a year during the war.

Many times have French War Ministers deplored to me their regrets that they could not have the span of Ministerial life which M. de Freycinet enjoyed, for without time to work out reforms little of permanent value can be accomplished. From 1888 to 1893 a succession of reforms increased the strength and the value of the army, and even when de Freycinet was not in office his advice was sought, for he was regarded as the inheritor of the Gambetta war traditions, and few men knew the army better.

His oratory did not compare, for eloquence, fire, and warmth, with some of the best of his day, but it was always most convincing, and there was a sweet reasonableness about it which was disarming. He was not a creator of scenes and incidents. By nature he was a man of science, and the power which he possessed in army affairs was due to knowledge of his subject, before which deliberative assemblies usually bow. The *Temps* has quoted in relation to him Gambetta's observation about himself, namely, that 'he would think he had stolen something from his country if he abstracted a single hour, or even a minute, from the cares of national defence, to devote it to internal politics.' There was, in fact, no French Minister more industrious, and it can truly be said of him that he never spared himself time or trouble, but from first to last gave the best of himself to his country.

Walther Rathenau

The death of Walther Rathenau causes Germany to mourn a great man rather than a great Foreign Minister, and we have every reason to regret that the bitterness of German politics has added another victim to the list of political assassinations.

Those of us who were in Berlin in 1920 and 1921 became aware that, after the Chancellor, Wirth, there was only one man in the German Government who really counted—namely, Walther Rathenau. Jew, socialist, magnate of industry, millionaire, prolific author, philosopher—there was scarcely any sphere of human activity, whether in the realm of thought or action, into which this man had not penetrated with his devouring and almost superhuman activity.

Son of the man who founded that great electrical combine, the A.E.G., he spent most of his early days either in University study or in the management and extension of his great industry. This work brought him into close relations with many other industries and gave him the almost encyclopædic knowledge of industrial affairs which he possessed. Though endowed with a manner which could not be called ingratiating, his brains compelled the respect of all. During the war he was called upon for public service and placed in charge of a newly-created service for the provision of raw materials. He was responsible for many of the Ersatz products with which Germany strove so hard, and for long so successfully, to annul the results of the blockade. When the peace came his place was clearly indicated in reconstruction, and it was he who met M. Loucheur at Wiesbaden last year to lay the foundation of payments by goods instead of gold.

It is one of the mysteries why the Chancellor called Rathenau to the German Foreign Office, but it is apparently true that the modern democracy supposes that a man who shines in one sphere must necessarily shine in another, and Chancellor Wirth was beset with too many anxious cares to take foreign affairs into his own hands. Rathenau was out of his element in diplomacy. Tall, *cassant*, and imperious, he looked more like an East Prussian Junker than anything else, and he showed no better acquaintance with the psychology of Germany's neighbours than most of his compatriots, though he knew several tongues. He was a type that might have gone far in philosophy or industry, but in diplomacy he displayed neither the experience, the prudence, nor the flexibility for the conduct of critical affairs.

In diplomatic history Rathenau will always remain connected with the Russo-German Treaty. He is credited with having been opposed to it, or at least to its signature, during the negotiations at Berlin. What particular reason impelled him to sign the Treaty at Rapallo on Easter Sunday during the Genoa Conference is still imperfectly known. Probably it was the dread lest the Allies should settle with Russia over Germany's head. In this case the hope probably was that the production of the Treaty would forestall the Allies and at the same time notably increase the prestige of the German Government at home. If this be the right reading, Rathenau's act amounted to a singular miscalculation. There was certainly credit given to the negotiators by the German Press, but this could not be of a lasting character if the price was the loss of Europe's confidence.

Rathenau's action showed him to be unmindful of the Cannes resolutions and ready to trick Europe for a German advantage. Europe had just admitted Germany to her counsels for the first time since the war on equal terms, and it was clearly of the first importance that she should seek to dispel the suspicions of the past. Rather, she confirmed them, and it is not clear that Germany has gained or will gain anything on the Russian side to make good this loss. The publication of the Treaty enormously aggravated the difficulties of Mr. Lloyd George in his dealings with Russia, since the Soviet became more intransigent in their attitude when they knew that they could fall back on Germany and on other separate agreements if a general settlement failed.

I happened to have been at Berlin, Wiesbaden, and Genoa when Rathenau was present. After the publication of the Treaty I was invited by letter to visit him at the Eden Hotel in Genoa, to listen to his explanations, and afterwards I had a private conversation with him. The explanation left me, and probably some others who were present, wholly unconvinced that Germany had done anything but a very stupid thing, even in her own interests. I asked, and others asked, a number of questions of the Minister, who answered courteously, but made his case no better. He asked me when we were alone afterwards what effect I thought the Treaty would produce in Europe. I told him that the effect would be deplorable. He alleged that he had tried to see Mr. Lloyd George for a week and had failed; that our Embassy in Berlin knew all about the Treaty, and that Baron von Maltzahn had on several occasions discussed it with two officials of the British delegation at Genoa. However this may be, the harm was done. Germany, having settled with Russia out of court, was thereafter excluded from the Commission dealing with Russia. The Chancellor and Dr. Rathenau had several conversations afterwards with the Prime Minister at his villa, but I have seen no public announcement of the result.

Walther Rathenau will hereafter be better regarded for his books than for his brief invasion of the sphere of diplomacy. As we read his books now, we ask ourselves whether it is possible for the manager of the A.E.G. and the author of the Russo-German Treaty to have been responsible for such writings, or inversely, we wonder how this prophet of a socialistic Utopia, this mystic, this idealist, can be identical with the millionaire magnate of industry and the German Foreign Minister whom we have known. Strange are the ramifications of the Teuton mind! If Lord

Haldane has a period of leisure, we should like him to take up that pile of Rathenau books and tell us what it all means. That Rathenau had ideas, and that he coined phrases that will live, we may acknowledge. But few English people will have either the inclination or the patience to persevere very far, and it is sufficient for present purposes to give a few extracts from his most popular work, 'Von kommenden Dingen,' in order to show in the dead statesman's own words how his mind worked.

'This book,' he says, 'treats of material things, but treats of them for the sake of the spirit. It treats of labour, want, and gain; of goods, rights, and power; of technical, economic, and political structure; but it neither postulates nor esteems these concepts as ends in themselves.

'The best among us grow weary of the work which only looks to to-day. For many, doubt, exhaustion, and despair become the central feature of their thought, so they give themselves up to the enjoyment of the passing hour, and renounce life's finest privilege—work.

'Others turn to dead dogmas and the promises of dead creeds. . . . But the faith which rests on dogma suffered from the futility of those temporal powers which were too weak to impose it upon the world, too weak to overthrow its competitors, and yet strong enough century after century to protect it with smoked glass from the radiance emanating from the peoples. This faith died when the protecting panes were shattered.

'To seek goals implies faith. . . . Because Socialism fights for institutions it remains at the level of politics. It can furnish criticism, can eradicate certain evils, can win rights. But it will never transform our earthly life, for the power to effect this transformation is given only to a philosophical outlook, a faith, a transcendental idea.

'In days to come there will be required something more than the renunciation of material goods. We shall have to put away from ourselves our dearest vanities, weaknesses, vices, and passions; upon us will be imposed the duty of cherishing sentiments and performing deeds which to-day we esteem in theory while despising them in practice; we shall have to learn by hard experience that our aim in life must not be happiness, but fulfilment, that we have to live not for our own sake, but for the sake of God.

'The blessedness of the Divine will interpenetrates us. Mankind will walk in this path amid enmity, scorn, and persecution. . . . Ingratitude will sanctify the path, hardship will accompany it.

'The man who lives solely for what he can filch from life . . . remains a hero only to one of his own kidney. . . . The arrogance of the superior person will never fertilise the soil. . . . Spiritual leadership will pass from women and smirking aesthetes to men, from poetasters and formalists to poets and thinkers.

'Genius is not the criterion of the soul, but the criterion of all creation is given in the awakening of the soul. . . . While our feet remain firmly planted on the earth our eyes must turn ever to the stars.'

CHAPTER X

Spain and Morocco

The Population and Armaments of Spain—The Spanish Protectorate in Northern Morocco—The Rifians—Population and Fighting Strength—Guns and Rifles—Abd-el-Krim—Probable Limits of His Influence—The Beni Urraguel—Pacification of the Eastern Rifians—Interference of Politicians the Main Obstacle to Spanish Success in the Past—Lines of Operation—Raisuli and the Jebala Tribes—Supposed History and Descent of the Rif Tribes—A Lecture by the Spanish Ambassador on the Country.

SPAIN, with a population of nearly 22,000,000 souls in her continental territory, has compulsory service, an annual contingent for this year fixed at 78,000 men, colour service for three years, 18 years of liability to service, and a peace establishment of 216,000 men, intended to present on mobilisation 16 active divisions, besides seven more found by reserve troops. The armament includes Mauser rifles and Schneider-Canet guns. Considerable efforts have been made of late to create an efficient air force. The Guardia Civil and Carabineros, constabulary, and military police, are additional to the army. In Morocco itself there are said to be about 150,000 men, of whom two-thirds are Spaniards. The navy, though not powerful, has ample means for succouring the army in any operations on the Rif coast, and is manned by 17,000 seamen and marines.

Such is the nominal position, but no one here can say precisely what the peace strength actually is, for what with the power to purchase exemptions from long periods of service and other inadmissible faults which have become conspicuous owing to favouritism and graft, we must regard all figures with a certain amount of friendly scepticism.

The Spanish Protectorate in Northern Morocco was delimitated by the Franco-Spanish Treaty signed at Madrid in 1912. It is roughly two hundred miles long and sixty in breadth, and extends from the Algerian frontier to the Atlantic. It holds a population estimated at over one million, including Berbers, Tuaregs, Arabs, Jews, and negroes. The Rifians occupy a large mountainous area in the centre and east of the zone from the river Muluya to the Wad Uringa, and their power is strongest to the south of the bay of Alhucemas. Out of a total population of 540,000 some 320,000 occupy the mountainous region. It is computed that the whole of the Rif tribes could, if united, place 50,000 fighting men in the field, but such union is problematical. Over the Beni Urraguel, Abd-el-Krim, whose village is near to the bay of Alhucemas, has been elected the 'spear,' or military head, and one adventurous journalist has recently penetrated to his headquarters. How far Abd-el-Krim's real influence extends we do not know, but, like the new 'King of Tirah,' he certainly cannot pay his warriors, and dissensions among his clans are common.

It is believed in Spain, and is probably true, that the particular band of stalwarts known as the Beni Urraguel are the head and front of the resistance to Spain, and that, could they and a few contiguous tribes usually, but not always, subservient to them, from fear or more worthy motives, be well beaten, the resistance would collapse.

Although Abd-el-Krim is the accredited and chief leader of this most warlike tribe, which is supposed to have 10,000 modern rifles at its disposal, much depends upon how far his real influence extends. The six Rif tribes in the eastern part of the Spanish zone, some of them very numerous, have been apparently pacified long since. Other tribes, to the south and west of Abd-el-Krim's local combination, are not always amenable to the nominal leader, and are often in open conflict with him. It is thought that only the Bekkoya—in Spanish, Bocoya—a smaller tribe also adjacent to Alhucemas Bay, are really inseparable from the Beni Urraguel, and that the Jensaman, Beni Tuzin, and part of the Beni Said—which latter has already in great part submitted to Spain—are likely to appear on the fields of battle only if they are bullied into alliance. Even in 1921, when the Rifs made their great effort, not more than 30,000 rifles were assembled; and it is possible that these numbers may not be exceeded unless recent gun-running has been unusually successful.

The Rifians are a strong and warlike Berber people, who have been a thorn in the side of every ruling dynasty in Morocco since the earliest times. They are very well armed with modern rifles, which they obtain from gun-running merchants, and have a few field guns taken from the Spaniards in 1922. But the highlands which they inhabit, and its inhospitable and roadless character, are their chief defence, and the intractable nature of its inhabitants makes an occupation indispensable if a permanent settlement is desired. This seems to be advisable while France is engaged in rounding up the Berbers in the Taza region and completing Marshal Lyautey's programme of pacification. Morocco is too near Europe and too rich for any part of it to remain closed for long to the unembarrassed trade of the world. The diminishing independence of Maghreb-el-Aksa draws near to its end.

The map shows the general disposition of the Spanish posts on the eastern side of the main Rif strongholds, besides other military details. The subjugation of the Rifians is a difficult problem, but one that is practicable, given ordinary preparations and precautions. The main obstacle in the past has been less the Rifian resistance than the interference of Spanish politicians at home, whose furious criticism of casualties and expenditure has had for its purpose the upsetting of any Government in office at the time. From this point of view the Spanish soldiers can well believe that the present situation at home offers them an unequalled opportunity for unimpeded action. If the Spanish air force is efficient, a new power is added on the Spanish side. But the task of the new Spanish leader, with a curiously Basque name, General Aizpuru, who has recently been sent to Melilla, will be mainly facilitated by the fact that he can count on being aided rather than impeded by Madrid. The separation which has been effected between the mountain blocks of the Rifs and those of the Jebala tribes on the west tends to facilitate the Spanish task.

There are many alternative ways of carrying an occupation of the Rif territory into effect. Since the disasters of 1922 the Spaniards have considerably extended and strengthened their hold on the eastern end of their zone. The map shows that they are not far from Annual, where the defeat of 1922 began. On the western side of the Rif country there is peace, and even the famous Si Ahmed ben Mohammed el Raisuli appears to be pacific, for what the appearance may be worth. This troublesome politician still exercises chief influence over the 21 Jebala tribes in the western part of the Spanish zone. He has been visited by General Aizpuru, and we can infer the object of the visit.

The Rif tribesmen can be attacked from the west, the east, or the south, or again from the sea, or some plan of gradually contracting a line round the tribes may be resorted to, combined with an extension of the railway system. It is surely a problem which the Spanish staff must have worked out very thoroughly, and they are not likely, in view of the past, to depreciate the work of a formidable enemy who keeps himself in good heart by persistently attacking the convoys which supply the Spanish advanced posts in the Melilla command.

It has been said that the gradual pacification of Gaul by the Romans was effected by a shrewd combination of terrifying, winning, and sharing; but the Rif Berbers will not lend themselves to be won over, nor are they ready to share anything with others. The origin of the Berbers is lost in the mists of antiquity, but tradition traces their descent from the great-grandson of Ham, the son of Noah, and they probably came from Libya or

Phœnicia at some indeterminate period, conquered the Maghreb, and threw back its negro inhabitants into the Sudan. When other conquering Powers came the Berbers retired to their mountains, and have there remained. Tradition says that they were twelve times converted to Mohammedanism by the sword and twelve times threw it off again. The Jebala tribes are slightly more orthodox, and keep the Ramadan fast.

The Berbers are in any case the foundation of the inhabitants of Morocco, and of their four groups the Rifians are the most untamable. They speak little Arabic in the Rif, and the prevailing speech is the Berber Kbad. In their savage highlands, which rise in places to an elevation of over 7,000ft, and are roadless, with narrow valleys between the hills, they are no mean foes. Three years ago the present Spanish Ambassador in London, his Excellency Señor Don A. Merry del Val, delivered a noteworthy address to the Royal Geographical Society on the Spanish zone in Morocco, and gave us the best description of the whole state of affairs there that has yet appeared. To that address we should turn if we desire to read a comprehensive account of the country and the people.

CHAPTER XI

The Unified Command

M. Painlevé's Disclosures—Political History and Origin of the Unified Command—Major-General Tasker H. Bliss's Views on the Evolution of that Command—M. Painlevé's Discussions With Mr. Lloyd George—Painlevé Becomes Prime Minister of France—Conversations—French Resistance—It is Overcome—Difficulties in London—An Accord Fifteen Days before Caporetto—The Two Premiers in Italy—The Agreement at Rapallo—The Paris Luncheon of November 12—Mr. Lloyd George's Speech—Painlevé's Government Falls—The Supreme Council—Painlevé the Author of the First Conception of the Unified Command—Want of Frankness of Mr. Lloyd George—Major-General Tasker H. Bliss's Revelations—The Versailles Military Representatives—Text of Sir Douglas Haig's Memorandum of March 2, 1918—Selected Indiscretions—The General Reserve—Claims Made by Bliss—The Reserve is Not Formed—The London Conference of March 14—Pétain's Assistance to the British After March 21—Why the Tide of Battle Turned.

I

M. Paul Painlevé, in the *Revue de Paris*, has made some interesting disclosures on the origin of the unified command in the war, and Major-General Tasker H. Bliss, in the American quarterly *Foreign Affairs*, has given his views on the evolution of that command. Both writers are competent authorities, for M. Painlevé was Prime Minister of France, in succession to M. Ribot, up to November 13, 1917; and General Bliss, after being Chief of the Staff of the United States Army, was American representative on the Versailles Military Executive set up by the Supreme War Council, when this body itself was created at the famous Conference at Rapallo in November 1917. In the articles named, M. Painlevé occupies himself chiefly with the political history and the origin of the unified command, while General Bliss is most concerned with the doings of the Versailles Military Executive, of which he was a distinguished member. The two stories therefore complete each other in a sense, and convey to us the ideas of two important people on a phase of the war which has been, and will be for a long time, a matter of much discussion.

M. Briand's Government fell on March 15, 1917, and M. Painlevé was War Minister in the Ribot Government which took its place. It was to M. Painlevé's merit that, after the defeat of General Nivelle in April, 1917, General Foch was rescued from a position of temporary inactivity and made Chief of the Staff in Paris and General Pétain appointed to succeed Nivelle in command of the French armies. As War Minister, M. Painlevé tells us that he had a long discussion with Mr. Lloyd George and Lord Milner on August 6, 1917, with the object of nominating Foch Chief of an Inter-Allied General Staff, and that on the following day it was decided that the Allied General Staffs should study the subject of such Inter-Allied Staff.

On September 13 M. Painlevé became Prime Minister of France, and tells us with perfect truth that the extension of the British line to relieve—*soulager*—the French, and the creation of better means for securing united action, were ardently desired by French opinion. M. Painlevé therefore met Mr. Lloyd George and had a private talk alone with him on September 25 in the Boulogne Station with the least possible publicity, and the two agreed upon a plan to create an Inter-Allied War Council, with an Inter-Allied Staff presided over by Foch as its permanent organ. Foch was to have under him the Franco-British reserves in rear of the point of junction of the two armies 'until the day when the state of opinion in England allowed him to be made generalissimo of the two armies.' As Mr. Lloyd George, on November 19, 1917, declared in Parliament that he was 'utterly opposed' to the appointment of a generalissimo, as it would 'produce real friction and might create prejudice, not merely between the armies, but between the nations and Governments,' it must be concluded that there is here a strange mystification which ought to be cleared up.

M. Painlevé, however, remained convinced that the unified command was to be carried through in two phases, in the first of which the indispensable organ of command was to be created, and Foch was to be charged, in a veiled manner, 'to co-ordinate the operations of the two armies' near their point of junction. M. Painlevé arranged with Mr. Lloyd George to go to London to discuss the extension of the British front and to settle the details of the other arrangements. Returning to Paris, M. Painlevé drafted the plan conformably, and in presenting it to the French War Committee explained that 'it was only a first stage to prepare the single command under General Foch as soon as British opinion was in a condition to accept it.'

The French War Committee was not easily won over. Some members asked whether the plan would not encroach upon the powers of the French Staff and upon the sovereign right of decision of each Government. What should be the powers of the new body? Was it not a third command added to two others, and therefore a new dispersion of wills rather than a concentration of them? By the help of MM. Franklin-Bouillon, Loucheur, and Doumer these objections were gradually overcome, and on October 8, 1917, M. Painlevé left for London with his project in his pocket.

In London he met with fresh difficulties. Mr. Lloyd George demanded that the articles should be more vague—*estompés* is the term used—and that if they were made precise they would have to be more restrictive. Three days of arduous discussion followed, but on Oct. 11 M. Painlevé left London with the decision that as soon as agreement was reached by the two Governments upon a common project it would be communicated as a provisional text to the Governments of Rome, Washington, and Petrograd, and that an inter-Allied meeting would be called, with the least possible delay, to reach a definite accord. M. Painlevé points out that this occurred fifteen days before Caporetto, and that it is an error to suppose that the Italian defeat brought about the plan. It was on October 10, M. Painlevé adds, that he also obtained the adhesion of Mr. Lloyd George to the extension of the British front, which was afterwards to have such serious consequences for us.

On October 27 the rupture of the Italian front occurred, and without delay Foch was sent to Italy, and divisions of French and British troops followed. Cadorna halted on the Piave and held on, while M. Painlevé hastened to London to complete the accord with Mr. Lloyd George on October 30. By November 1 the accord was completed; the two Premiers left for Italy on November 3, and on November 5 began the discussion with MM. Orlando and Sonnino at Rapallo. The pact was completed and signed on November 7, and Mr. Lloyd George insisted that Foch should no longer be Chief of the Staff of the French army at Paris, but at the head of the military executive at Versailles, that is to say, a councillor at the disposal of the British Government as much as of the French.

M. Painlevé returned to Paris on November 10 when he became aware of the fall of Kerensky's Government and the hopeless decomposition of the Russian army. He therefore organised the famous luncheon of November 12 to affirm in a striking manner the unity and determination of the Western Allies. In a speech he alluded to the Supreme Council created at Rapallo, and spoke of 'one front, one army, and one nation' as the programme needed for victory. Mr. Lloyd George, in a speech which M. Painlevé declares had been written on the shores of the Mediterranean, defined the military executive of the Supreme Council as 'a stage which must be completed before going farther,' and made an appeal for united action, but M. Painlevé omits reference to the passages in the speech which gave such deep offence to our Army Chiefs in France, though he acknowledges that the Rapallo Pact was ill-received in England.

In the Chambers he had a hard task, owing to the French scandals which created such distrust at this time, and on November 13 his Government was overthrown. But in the debate he pointed out that 'the working of the Supreme Council will create, in fact, without saying it, this unity of command, which is worth more than having the name without the thing.' He tells us that this is all he could say in reply to M. Abel Ferry's harassing question, 'Yes or no; is General Foch the supreme chief?' To this brutal interrogation, says M. Painlevé, he could not give a similar reply 'without compromising Mr. Lloyd George's position in the House of Commons.'

M. Painlevé is a very trustworthy witness, and, failing any refutation of his account, must be set down as the author of this first conception of the unified command. But it is strange that Mr. Lloyd George's secret approval of the Generalissimo plan, which is implied throughout the article, should be in contradiction with the late Prime Minister's statement to Parliament, and that the whole of the arrangements of the two Premiers between August 6, 1917, and November 5 should have remained concealed from the British General Staff. This seems to follow from Field-Marshal Sir William Robertson's book, wherein it is stated on p. 327 that the Allied military authorities discussed a new system of command in the summer of 1917, 'but again nothing came of the idea.' In the very next paragraph the Field-Marshal says that 'the question next came up at the Rapallo Conference,' from which we must assume that he was not informed of the purport of the discussions between the two Prime Ministers on September 25, or October 11, or November 1, nor had any idea of the aim of the two Prime Ministers to nominate an Allied Generalissimo, which is not indeed strange, since the late Prime Minister expressly denied that it was his aim.

This most unfortunate want of frankness, the attacks of part of the British Press on the General Staff which began, unchecked by the censorship, towards the end of October, 1917, and the failure to define what was meant by the word Generalissimo, were largely the causes of dissensions which followed. Had Mr. Lloyd George told the General Staff that he wanted a single French commander to control all the armies on the Western front, my impression is that the General Staff would have agreed with him, and would have worked out a far better plan than that which was imposed before Doullens and Beauvais. The confessed object of the Rapallo pact was the 'better co-ordination of military action on the Western front,' but the organ set up was political, and could not improve military action, because the Military Executive at Versailles was, as the French War Council had warned M. Poincaré, only one more commander added to the others, while it set up a body of soldiers to advice the Supreme Council over the heads of their General Staffs, who had the duty of giving their Governments advice on strategy. The result of this hybrid plan, which was opposed to all sound principles of military organisation, will be shown in an examination of the article of General Bliss.

II

The American public, we must admit, are privileged. General Bliss gives them his story of the Versailles Military Executive and of the famous four days' conference of the Supreme Council at Versailles, January 30—February 2, 1918. He accompanies it not only by extracts from his own official memoranda, but by quotations from selected speeches, by textual statements of resolutions hitherto carefully concealed, by a reproduction of General Foch's map showing his idea of the distribution of the General Reserve in anticipation of the German attack, and even by the text of Haig's memorandum of March 2 regretting his inability to comply with the suggestion of the Versailles Board that he should detail nine or ten divisions to form part of this General Reserve. I do not recall that I have ever seen before some of these papers, and the text of Haig's memorandum is not given in Mr. Dewar's Chapter iv., Vol. II., of 'Sir Douglas Haig's Command.'

It is really admissible to grumble just a little. These partial indiscretions, however interesting and valuable, whet the appetite for more. We are most of us in favour of giving complete publicity to the whole of these and kindred documents, but for an American general to select certain secret papers and publish them when the rest of the people associated with America in the war have not the whole story under their eyes, borders on the vexatious. With all respect to General Bliss, his extracts are certainly selected. He tells us himself that the Versailles soldiers submitted 'some fourteen Joint Notes' to the Supreme Council on January 30, but he only gives extracts from Nos. 1 and 12, relating to a General Reserve, and we are left in blissful ignorance of the other twelve joint notes, including that on the Turkish campaign. But we know, or believe we know, that the Eastern offensive was adopted by the Council on the understanding that no forces should be removed from the West to take part in it, which decision did not allow our General Staff to bring to France, until after our disasters of March 21, the British troops engaged in the East.

However, the main object of General Bliss is to prove that, had the General Reserve been created, the disaster of March 21 would have been avoided. That is precisely what he does not prove, nor can prove. 'No one can now deny,' says the American general, 'that Foch's plan was the only one that could have assured prompt assistance the moment the crisis developed.' But this can still be denied, and is disproved by the evidence of facts. The creation of a General Reserve was, in principle, settled by Resolution No. 13, passed by the Supreme Council on February 2, the text of which—and the powers of the Versailles Military Executive in connection with it—are given verbatim by

General Bliss. The Versailles soldiers promptly and properly set to work to create the Reserve, and settled that it should consist of about 30 divisions, namely 9 or 10 British, of which 3 were temporarily in Italy ; 13 to 14 French, of which 4 were in Italy ; and 7 Italian. It was proposed that these divisions should be kept in the zone of the armies to which they belonged, and Foch's little map, which General Bliss reproduces, shows the British contingent of the prospective General Reserve scattered round Amiens, from the level of Arras to that of St. Quentin, and the French contingent scattered over an immense front between Troyes and Paris. As the Italian divisions were to remain in Italy the actual number of Franco-British divisions in the proposed General Reserve would have been 16 to 17.

What was the conception of the Versailles soldiers in making these proposed distributions ? Here was the opportunity for General Bliss to reproduce, with a map, the famous prophesy of Versailles respecting the coming German assault. There is not one word about it, and report says that the attack was not expected by Versailles till the summer, and then in a different sector of the front from that where it actually occurred.

However, the General Reserve never materialised. Haig was unable to supply the divisions, so were the Italians, and the French contingent was whittled down to eight divisions excluding the troops in Italy. In consequence, Foch's committee found itself on March 4 unable to continue its work, and each military representative was asked to report the fact to his Government. General Bliss wrote to his committee that the instructions of the Supreme Council were mandatory, and that Versailles could not abdicate its duties ; but when the Supreme Council met in London, on March 14, Mr. Lloyd George upheld Sir Douglas Haig, and M. Clemenceau closed the account by stating that 'for the moment it is impossible to withdraw divisions from the command of Field-Marshal Haig and General Pétain in view of the threatened attack.' All that the Supreme Council could do for doleful

Versailles was to suggest that a committee of four general officers should proceed to Italy to see that the French and British divisions in that country, and a quota of Italian divisions, should form the nucleus of a General Reserve. But the mission had no result, and had scarcely returned to Versailles before the German attack was launched.

Sir Douglas Haig's memorandum on the proposed abstraction of his six or sëven divisions to form a General Reserve is evidently resented by General Bliss, but the British Commander's reasons were unanswerable. He said that an offensive appeared to be imminent, that all his troops were disposed to meet it, and that, were he to withdraw the divisions demanded, the whole of his plans and dispositions would have to be remodelled. He also foresaw—and correctly foresaw—a wider employment of Allied Reserves than that foreshadowed in the Versailles Note of February 6. He said that a sustained hostile attack in great strength might entail the despatch of a considerable force to the assistance of the army attacked, and that such force could not be ear-marked or located in any particular area prior to the delivery of the attack and the development of the enemy's intentions, since the situation might demand the employment of the whole of the resources of one army. He added that the Franco-British arrangements for mutual support had been made, and he gave some details of them.

Now, if we follow out General Bliss's ideas, which are apparently those which prevailed at Versailles at the time, we see from Haig's despatches that actually 8 British divisions were drawn into the battles of the Third and Fifth Armies, over and above all the local and general reserves of these armies, in the days following March 21; while the French, instead of supporting us with 8 or 10, or even 14 divisions, had 16 divisions on the battlefield by March 26, and 27 more on the way there. So that when General Bliss speaks of the 'futile attempt' of Haig and Pétain to provide an Allied reserve we can only say that it was the reverse of futile, and that it was stronger by 21 divisions than the theoretical Foch reserve at its unattainable maximum, and was probably available in a shorter time. A nice mess we should have been in when the theoretical Foch reserve had been exhausted and Foch had striven to collect more from Armies not under his control! One would imagine from General Bliss's account that Haig and Pétain did not succeed in arresting the German attack. Yet they did so, and General Pershing, in his final report, says that 'only the prompt co-operation of the French and British G.H.Q. stemmed the tide.'

It is surely very bad teaching for any general at this time of day to exalt such a monstrosity as a General Reserve under a different command from the fighting armies. If Foch had had the powers subsequently entrusted to him at Doullens and Beauvais, it would have been a totally different matter, but what was so bitterly resented was the semi-civilian plan of removing from the Commanders-in-Chief reserves which came within the sphere of the battle, and placing them under the control of some-one else, an amateur conception entirely opposed to sense, reason, and military experience. We laud, with General Bliss, the 'good sense, kindly tact, personal magnetism, and supreme professional qualities of General Foch,' which had a proper sphere of usefulness when, at Beauvais, on April 3, he was given the duty of 'co-ordinating the action of the Allied armies on the Western front,' but because his name was connected with the General Reserve of the Versailles scheme we are not compelled to abandon our right of private judgment or to call a scheme good when we know it was bad.

It is, of course, quite a mistake to imagine that the mere appearance of Foch as virtually the Allied Commander-in-Chief on the French front made everything at once go as happily as marriage bells. Between March 26 or April 3 and July 18 there was a fearful struggle, and two months after Doullens the French armies met with a great defeat, comparable with ours of March 21, though Foch had had for those two months the entire strategical direction of the Allied armies. We were in presence of the supreme German effort, and in presence, also, of a really superb tactical handling of the hostile mass attack combined with surprise on the part of our enemy.

If we want to know why the tide turned in our favour we should remember that only 300,000 American troops were in France by March 21, and that in July these numbers had increased to 1,200,000. Between March 21 and the Armistice no fewer than 740,000 of our own men reinforced the British Armies in France, and the French armies also kept up a strength which was not at one time thought possible owing to the use of native troops on a large scale. The brilliant generalship of Foch could do nothing till the tide of numbers turned in his favour, no more than Napoleon himself could do when the same tide turned against him. Foch himself, by apparently consenting to such a ridiculous plan as a General Reserve detached from the Command of the Western armies, must, I imagine, have had his tongue in his cheek. He probably knew the object at which M. Painlevé and Mr. Lloyd George were aiming, and was waiting, as they were, for opinion to ripen. But it was all a mischievous aud dangerous game, for had we been attacked with the Foch General Reserve in being we should probably have been in Queer Street.

CHAPTER XII

The High Command in Coalition Armies

General Debeney's Study—Campaigns Between 1813 and 1914—Conditions of the Realisation of the Single Command—Critical Events and a Strong Personality—The Formula for the Single Command of the Central Powers in the World War Not Found Till September, 1916—A Tremendous Business to Improvise the Entente Organisation—The Entente Never Realised the Single Command of the Coalition—Only Unity of Direction in the Western Theatre—General Debeney on Our Future Problem—Military Conventions and Inter-Allied Staffs—The Personality of the Commander Emerges from Events—An Appeal for the Spirit of Coalition—Comments.

An Historical Survey

General Debeney, the distinguished Director of the famous French Ecole Supérieure de la Guerre up to the moment when he became Chief of Staff of the French Army, recently gave a lecture on the question of the command of armies of coalitions—a subject to which we seem to have paid little or no attention since the war came to an end. It is worth while to summarise the General's views from the text published by the *Opinion* and to make a few observations upon the lecture.

He began by declaring that 'The World War' is a correct term to use, because that was the distinctive character of the last war, and it is almost certain that a conflict between two Great Powers will not easily be localised in future, and will tend to bring in other States, owing to the community of interests of the world. Therefore, it is our duty to study the question of the command of these coalesced armies which we must foresee. As the French rule extends over 100 millions of people, and the British over 450 millions, it is no light task to organise the higher command of their united armies. By an investigation of coalitions since 1813, the General suggests that we can determine how the problem presented itself in the past, and, in the light of this experience, how it will present itself in the future.

So he glanced at the conditions in former campaigns, and showed that in 1813 and 1814 the principle of unity of military command against Napoleon was accepted, but the difficulty was to find the man to lead. He regards Schwarzenberg as a man of mediocre talents and inclined to middle courses, but says that he was the candidate of the most adroit diplomatist of the day, namely, Metternich, and that the appointment overcame the difficulty of choosing either the Tsar Alexander or Blücher, both of whose characters, in different ways, were somewhat inconvenient. It was not till the close of 1813 that the command was definitely confided to Schwarzenberg by a War Council at Frankfort, but even then the command proved to be nominal, because each event provoked a new meeting of the War Council, at which Sovereigns and Ministers were present. In effect, the various national armies each went their own way, and it was only the menace of a great defeat that drew them together.

General Debeney thinks that in 1815 unity of command was not formally agreed to, but was, in fact, exercised thanks to Wellington's strong personality, which extended over policy as well as strategy. It was he, General Debeney thinks, who really conducted the war, even if the pompous title of Commander-in-Chief of the Austrian and Allied Armies was still attributed to Schwarzenberg. In fact, a study of the great coalitions at the opening of the nineteenth century shows General Debeney that, to realise the single command effectively, two conditions are necessary, namely, highly critical events and a strong personality.

General Debeney shows that during the Crimean War there was no single command, and that the Governments, in the Treaty of March 12, 1854, only stipulated that 'the general plan of campaign will be discussed and determined between the Commanders-in-Chief of the three armies.' So personality had the casting vote, that of Raglan at the beginning and of Pélissier at the end, but General Debeney considers it fortunate that the coalition was put to no further strain after the fall of Sebastopol. He thinks that the good comradeship of the Crimea ripened into the greater comradeship of the World War, and was to some extent the explanation of it. He then drew deductions from the wars of 1859, 1864, 1866, 1870, and the Balkan campaigns, and specially notes the Prussian disinclination to disclose plans to allies in 1866, and the tendency to impose personal plans upon other allied forces, disclosing a mentality which reproduced itself in 1914.

We generally regard the Coalition of the Central Powers in 1914-18 as a feudal arrangement, in which the Kaiser's allies played the part of vassals. This was true of Bulgaria and Turkey, but General Debeney observes that it was never true of Austria, and that the memoirs published since the war leave no doubt on this subject. Falkenhayn has written that the question of the conduct of war by a coalition had not been settled either before the war or when it broke out. Ludendorff is in accord, and declares that Berlin feared to entrust its secrets to Vienna. So the fault of 1866 was repeated, and the price this time had to be paid. Had the Prussian plan against France succeeded, no doubt the march against Russia would have then begun with great prestige, but much of this was left on the banks of the Marne, and the relations of Conrad and Falkenhayn were devoid of sympathy. Falkenhayn's dismissal, the fearful losses of the Germans at Verdun and on the Somme, the check at Asiago, the defeat on the Isonzo, and the entry of Rumania into the war, still further lowered the prestige of the German General Staff amongst its Austrian allies.

It was when this critical situation arose, and only then, after long and difficult negotiations, that the formula for the single command was found and adopted (September 6, 1916). The Kaiser then had the supreme command, and the directives were given by the Chief of Staff of the German field armies to Germany's allies. The result was good at first, but political rivalries about Poland deeply affected events, and General Debeney argued that the great political crime of the eighteenth century was far from being without influence upon the fall of the two empires which had despoiled Poland.

On the side of the Entente, General Debeney thinks that the Franco-Russian Alliance and the Anglo-French Entente were defensive instruments only, and were therefore subject to the diplomatic offensive of the Central Empires—that is to say, to their strategic initiative. But relations had been established, and besides our cause was just. To this we must mainly ascribe the successive entry of numerous Powers into the war on our side and our final success. But it was a tremendous business to improvise the organisation of the vast resources of the Entente ; and when there was a cry for the single command it was a cry more easily raised than realised. It was arduous, says the general, to adjust in a single theatre the action of several large armies of different nationalities. It was more arduous to combine the operations of armies and navies. But the difficulties reached a climax when it became a question of co-ordinating operations not only on land and sea but over theatres separated by thousands of miles in France, Italy, Macedonia, Russia, Mesopotamia, Palestine, Africa, and in all the different seas of the world.

As the Governments gradually began to understand the difficulties things began to mend, but never reached the stage of the single command of the coalition, though General Debeney thinks they might have done so had the war been prolonged. So far as things went, he says, we reached the stage of 'unity of direction' in the Western theatre—France and Belgium—only ; and General Debeney observes that the term 'unity of command,' even in this theatre, cannot properly be used. Once more, he declares, it needed both the crisis and the man even to get so far.

It remains to give the lessons which the General draws for future application, and to make a few observations thereon. Our problem, he says, is to create unity of command in coalition armies of the future and to prepare the inter-allied organisation of economic resources. We cannot rest content with the organisation of each national force separately. The single command of military forces can be realised by military conventions and by the preparation of inter-allied staffs. Military and naval conventions have capital importance, but are a corollary of an alliance or at least an entente. They lead to combined studies by the respective staffs and pave the way for an inter-allied staff, but it is visionary to expect that the man to lead can be chosen in advance. The personality of the commander will emerge from the events. The General holds that the basis of all effective co-operation is the creation of an inter-allied staff and that the men to form it must be trained in advance, instructed in the necessary foreign languages, and acquainted with the manners and customs, as well as with the armies of the nations likely to take part in the coalition. In the economic sphere it is not possible in peace time to realise the centralisation required in war, but what can be done is to foresee the organs indispensable in war-time, to prepare measures and regulations, and, in short, to make ready against the future without present dislocations and disturbances.

This economic mobilisation is essentially an act of Government, says the General, and he takes credit to his own country for having created the necessary organ to deal with it—namely, the Conseil Supérieure de la Défense Nationale, assisted by the Commission d'Etudes and a permanent secretariat. A Bill on national mobilisation is promised as the result of the deliberations of these bodies, and General Debeney inclines to the view that a law of this sort will confer upon France special facilities for joining a modern coalition. Finally, he appeals for an 'esprit de coalition' which will prepare the education of public opinion by appealing to moral forces and to the community of ideas and sentiments which the existence of a coalition demands.

We have not yet reached the stage of settling the relations between the different parts of our own national forces, but it is obvious that the problem which the late head of the French Ecole de la Guerre presents to us is in the same general order of ideas as those which our Defence Committee have recently been in process of examining. This body might well make sure that any new national system fits into the ideas which the French general discusses. In one way or another we must have a definite conception of how we intend to co-operate with an ally or allies, and this is a subject upon which our respective staffs and staff colleges may fruitfully confer.

We must all agree that in a coalition there are no suzerains or vassals, but only equal States which may be prepared in certain eventualities to confer upon some leader of marked distinction the power of directing, locally or generally, allied armies and fleets, or both, for the common profit. The tendency will be for the direction to fall to the Power which, in any particular theatre, employs the largest forces on sea or land, and no one can say which Power will be in this position in advance, since this will depend upon circumstances, upon the theatre of war, and upon the grouping of coalitions. Therefore each Power must be prepared both to direct and to be directed, and such an incident as Chanak must not recur when we are all in line upon general principles.

But it is difficult, as the last war and all history for a century past have shown, to get Allied Armies and their Governments to accept one 'co-ordinator,' even in a single theatre on land, while at sea we English never obtained the single direction to which we thought ourselves entitled in 1914-18. It is true, and it will probably remain true, that only highly critical events and a strong personality produce the single direction. Even if it comes we must recognise that the powers even of a 'co-ordinator' in a single theatre are dependent upon his success, for a Government which sees its armies or fleets defeated under a foreign director will cry out. The single direction is little fitted for a war in which a series of defeats occur, and in this respect it is inferior to the old practice.

An actual command of our armies and fleets by a foreigner would be impracticable. The directive may come from a foreigner, but our own people must be commanded by their own leaders. We think, and rightly think, that the directive of Marshal Foch led to our triumphs of 1918. So they did, but it is necessary to recall that from March 26 to July 15 they produced defeats. To understand the insecurity of tenure of a Marshal in such a position it is necessary to have before us the correspondence of Governments during that period, and this we have not yet been given.

But, of course, there is something more, for the general lines of the strategy of the coalition must be laid down at the outset of a war by the respective Governments in consultation. Changes may take place subsequently by mutual agreement, but there are usually a principal theatre and secondary theatres, that is to say, one in which victory will be decisive and others in which victories or defeats will be indecisive. Nothing atones for a false direction given to the main effort at the outset, and the preliminary decision is far more important than any other. Discussions between naval and military staffs in peace may or may not have foreseen in advance the general conditions and the grouping of the combatants, but political and other considerations may intervene and result in some fatal compromise. It is with anguish that we read even now of the extraordinary Council in London on August 5, 1914. There we saw debated as an open question the arrangement agreed upon by our staff and the French as far back as 1906 to meet the eventuality which actually occurred.

No doubt inter-Allied staffs, properly trained and in constant touch, may possess a common doctrine, but it will be necessary to define what General Debeney means by his inter-Allied staff. I have an impression that he implies a single staff which will more or less control the conduct of the whole war. That would be ideal if all Governments would abdicate their functions and obey the orders of the inter-Allied staff, but such conditions are not to be expected. Nothing which neglects the modern democratic ideas and the responsibility of national Governments towards their respective people is worth pursuing. So I wish that General Debeney would give us a second lecture on the constitution and functions of his inter-Allied staff.

However, General Debeney has given us something to think about, and in due time, by a course of investigations and discussions, the right solution may be found. If we are so cursed as to be faced by fresh wars they will probably be wars of coalitions. All the implications of such events must be considered, and we must endeavour to transmit into our exceedingly rebellious British hearts that *esprit de coalition* for which the French general pleads.

CHAPTER XIII

The Co-ordination of the Defence Forces

THE DEFENCE COMMITTEE*

The Report of the Committee of 1923—The Treasury Minute of 1904—The New Chairman of the Committee of Imperial Defence—A Keeper of the Prime Minister's Conscience in Defence—His Duties—A Super-Chief of a War Staff in Commission—An Advance upon the Path of Progress—The Controversy between the Admiralty and the Air Ministry—The Task of the Air Ministry—It Must Have Complete Liberty of Action—Home Defence the Primary Function—Campaigns in which the Air Forces Become an Auxiliary Arm—Excessive Claims of the Air Minister—The Field Service Regulations.

MR. BALDWIN'S Government decided to accept the recommendations of the Committee which are to be found in the statements made to Parliament on August 2, 1923, and in Command Paper 1938.†

Though the relations between the Air Force and the Navy attract the most attention and arouse somewhat intense feeling, the question of the future constitution of the Committee of Imperial Defence, and of its composition, aims, and methods of work, transcends all others in importance, and must be considered first.

It is a silent but eloquent tribute to Lord Balfour that the Treasury Minute of 1904, which created the Defence Committee, should have remained almost unchanged until to-day, and should still remain in principle unaltered. Experience, new ideas, and

* *See* Appendix I.
† *See* Appendix I. and Appendix II.

new arms, require some amendment of the processes of considering defence, and these have been taken into account in the changes effected by the Government ; but substantially the Defence Committee remains what it always was, namely, a consultative and advisory body, created to supply the Prime Minister and his Cabinet with the best available advice on all matters connected with defence.

The final report, Cd. 2029 (Appendix II.), gives the reasons why a Minister of Defence, and alternatively an amalgamation of the fighting Services, have been rejected. I have consistently opposed the plan of a single Defence Minister, and the final report recalls the many and weighty reasons against that solution of our difficulties. But it is true that the action of the Defence Committee has hitherto hinged mainly, if not exclusively, upon the Prime Minister, who, with his constantly expanding responsibilities, or it may be with his distaste for military matters, might be incapable of giving that time to defence which its importance and its cost demand. In this event a consistent line of policy would not necessarily be preserved. Again, there has been the fact that, though Departments have had some powers of initiative, actually the Prime Minister's conscience has had to be conquered before anything serious could be done. The Defence Committee has hitherto been exclusively the Prime Minister's Committee, and he has not only decided what questions should be examined, but has had the privilege of deciding what members should compose the Committee in each succeeding investigation, and even what witnesses should be called before it.

It is not true to say that the Prime Minister has surrendered any of his rights. He remains President of the Defence Committee, and can preside over it when he pleases, and summon anyone he likes to council. But for the unending spade-work of investigation he has seen fit to be aided in future by a Deputy who will be chairman, and will preside over the Committee in the Prime Minister's absence. In place of a somewhat vague and uncertain membership, he has placed on record the Ministers who should be permanently members of the Committee ex-officio, and has added to them the Chiefs of Staff of the three fighting Services. Dominion Ministers can be invited, and any other officials or persons can still be summoned as members by the President.

The chairman thus becomes the keeper of the Prime Minister's conscience, and of his memory, in relation to defence, and his personality, in successive Governments, will always be of immense consequence. The type is a man like Mr. Asquith—that is to say, a man who to vast governmental experience adds a cool judgment and a judicial mind. Almost necessarily he should not be burdened with the administrative toil of a great department, and should be able to devote his whole attention to defence, which is a subject large enough to attract a first-class mind, since it extends with indeterminate limits in every direction.

The chairman not only presides over the Committee in the absence of the Prime Minister ; he reports to the latter and to the Cabinet any recommendations and interprets any decisions thereupon made to the departments concerned. He has also the extremely important duty of keeping the whole situation of defence, and expenditure thereon, co-ordinated and framed to meet policy, and makes such investigations that full information on all matters concerning defence may always be available for the Committee. For this purpose the three Chiefs of Staff are placed at his disposal. They are to have individual and collective responsibility for their advice on defence, and these three will constitute ' a Super-Chief of a War Staff in Commission.' They are to meet together, possibly at the suggestion of any one of them, to discuss questions concerning their joint responsibility, and are apparently intended to agree upon recommendations, or to agree to differ, before they meet the chairman.

Questions relating to the co-ordination of expenditure may be referred to the Committee by the Cabinet, and will be considered in relation to the general defence policy of the Government and to strategical plans drawn up to give effect to that policy in war. A few minor improvements, such as the strengthening of the staff of the Committee, conclude these very interesting announcements.

These arrangements appear to be very sane and sound solutions of a difficult problem of modern statecraft, and to allow full weight to policy, expenditure, and the opinions of the fighting Services. They represent more nearly a real co-ordination of effort than anything in the same order of ideas that has preceded them, and are far more practical and sensible than any other proposals yet made. It is obviously expected, or at least hoped, that the three Chiefs of Staff will agree amongst themselves, and will seek to harmonise any conflicting opinions that they may hold, before meeting the chairman in council. It will be the chairman's duty to bring all three into line when they do not agree. They can all explain fully and frankly to the chairman their military opinions, and not sit silent as they frequently have to do in full Committee, while their political chiefs endeavour to expound things which they may not understand.

But the Chiefs of Staff evidently can neither supersede nor override their departmental political chiefs in any way, for everything has to come up before the Committee in the end, and there the First Lord, War Secretary, and Air Minister will be present. The political chief of a fighting service is in truth an administrator and nothing more, and the sooner that aspiring administrators abandon the habit of thinking themselves the directors in war of armies, fleets, and air forces the better for all concerned. Each man to his task, and the work is well done.

I have already described the French system of war preparation and the duties of the French National Defence Council. From some points of view the French system appears over-organised, and I prefer our own. But it must be admitted that, with regard to the general organisation of the people in time of war, victualling, transport, and manufacture, our system makes no apparent provision at all, while none of the ex-officio departmental heads, now made members of the Defence Committee, are concerned, or will be concerned, in these most serious duties in a great war. In fact, we have not yet decentralised control nor have we brought in all departments of State to take their part in a general mobilisation. This may come. The new Defence Committee may adapt itself hereafter to the recognition of these great war needs. But at present it does not, and limits co-ordination to the functions of the Cabinet, the Treasury, and the fighting Services. Therefore this latest reform, valuable though it be, is only a stage upon a long road still in front of us.

Navy and Air

The controversy between the Admiralty and the Air Ministry, as well as between their respective supporters in Parliament and the Press, concerning the control of the Naval Air Service, has been a serious one, and the general public have been bewildered by rival claims and apparently unanswerable arguments brought forward by the two parties to the dispute. On the one side stands the Navy, our senior fighting Service, endeared to the people of this country by the memory of wars and victories innumerable, and trusted still as the 'sure shield' which is held between England and her assailants. On the other side is a junior Service, born but yesterday, brilliant in its first campaign, appealing particularly to the imagination by the conquest of a new element, and suggesting future progress almost beyond the mental grasp of the average man. Neither Service was able in the World War to defend our shores and our cities from the ravages of an enemy arriving by way of the air, nor to prevent the infliction of immense loss upon our sea-borne trade, but the Air Service played a valiant part in overcoming the enemy's airmen in France, while the Navy carried out its main mission and was a potent instrument in causing the ultimate exhaustion of the enemy.

At the present moment the creation of air forces equal to those of the strongest air Power within raiding distance of our shores is of greater importance than the question which service shall control these forces when they are created. The creation of these necessary air forces is a work of such immense complexity and importance that the Air Ministry must be granted complete liberty to achieve it, and not be delayed or hampered by departmental rivalries, which exhaust energies and occupy valuable time. On these grounds alone the decision of the Government to preserve the unity of our air forces is, in my opinion, incontestable.

Before we make up our minds upon the uses of our air forces in war we must examine what character of campaigns these forces must be ready to meet. These campaigns vary infinitely from the maximum of national war against an enemy within air-raiding distance of us to some minor campaign of a colonial character. Whereas all wars of less importance than a national war against an air Power near to us are usually not an immediate menace to the existence of this citadel of the Empire, a national war of the above character is such a menace, and, moreover, it demands that all our available air forces at home should be ready to act with decision at the actual moment of the declaration of hostilities, since he who is first at work in the air gains a supreme advantage.

It is therefore necessary that home defence in the air should be the primary function of our air forces at home, and that the Chief of the Air Staff, and not the Air Council, should have the direction and control of them in his own hands. This control he can have under the recent decision, and he can agree beforehand with naval commanders upon the fractions detailed for duties with the Navy. Such forces, once allotted, cannot, under the recent compromise, be taken from the Navy except by a Cabinet decision.

But this case of a great national war against a strong air Power near to us is not the only form of war to be provided for.

In nine cases out of ten, judging by the past and taking count of the present, wars likely to occur will be those in which the Air Force will not necessarily be more than an auxiliary arm for the Navy or the Army. In the initial stages of a war in the Pacific, for example, the Naval Commander-in-Chief should control all the air forces under him, subject to consultation with the Air Force Commander on his staff, and in the same way, in a war on land, the Military Commander-in-Chief must direct his air forces, which should be part of his army.

It is necessary to point this out because even the Air Minister, Sir Samuel Hoare,on May 1, 1923, spoke of the Air Force as 'a primary and independent arm with its own strategy,' and admitted that he held this view. In a national war of the character described above we can admit the claim for home defence,subject to the policy of the Government and the organised co-operation of the other Services. But the extension of this claim to all wars is inadmissible, for in our distant wars one Service or another predominates, and the others take second or third place. We saw the germ of a wholly indefensible idea when in 1918 the Air Ministry was weakly permitted to direct its 'independent' squadrons in the east of France, which were not under the command of our G.H.Q. Had they been so, August 8 might have been a German rout, and I found no one at the time, outside the Air Ministry, to uphold such a senseless proceeding.

In spite of this experience, the Air Force has been allowed to perpetuate its inadmissible claims. In the first volume of the Army Field Service Regulations, only published recently, we read in Ch. XII., Sec. 76 (1) that "the Air Council may in time of war provide an air force in any given theatre, under an air officer commanding, to carry out such operations as the Air Council may from time to time direct in pursuance of the war plans of His Majesty's Government." The amendment needed here is that the Air Council may not do any such thing, for to suppose that the Air Council in London can fight its own campaign on the

North-West Frontier of India, while the army may be fighting a different campaign, is against all sense and reason. If the Air Council makes similar claims in a naval war, say in the Pacific, it will be the best way to lose a war ever invented by the wit of man, and the sooner that the new chairman of the Defence Committee looks into this matter the better. So far as I recall, no member of the present Air Council has ever set foot in India or the Pacific, and yet they claim the direction of strategy there! It is these preposterous claims which withhold from the Air Council the complete confidence of the other services and the public.

CHAPTER XIV

Imperial Defence

General Considerations

General Considerations—Policy and the Services—Position of the Dominions and of India—Commitments and Risks—Mandates—Policy and Armaments—Wars and Needs—Concentration *versus* Decentralisation—Three Kinds of War—No Preparations Made for a National War—Food Supplies—Navy and Air—Comparative Air Strengths—Abolition or Equality—The Army Needed—The 1914 Scale for the Regular Army Should be Restored—Record of the Coalition Government of 1918-22—The Responsibility for the Army Reductions of 1922.

THOSE among us who have spent much time since the war in visiting Europe and America to make ourselves practically acquainted with the changes resulting from the great events of 1914-18 returned home with a very natural desire to know what the Governments of Great Britain, the Dominions, and India had been doing all this time to place the Empire in a position of security, to maintain treaties, and to preserve peace, which remains now as ever the first of British interests.

We were considerably disillusioned on turning over in our minds the distracted state of the world and comparing it with our defensive arrangements so far as they had been announced. The state of public finance is no secret, but other nations have acted on the principle of 'safety first,' and we have not. We seem to be the only State in the world, not even excepting the United States, that believes in peace; the only State that has reduced its peace establishments to the danger-point even for the internal security of the Empire ; and the only State which has made no serious provision for the recurrence of a national war, even by the most superficial arrangements. Finance and politics rather than facts and foresight have been our guiding lights. We have not only jettisoned the armed forces that won the World War, but have inexcusably neglected the lessons which that war taught—lessons which should have become the inalienable inheritance of the Empire, and should have been applied long ago to the frustration of designs which may imperil peace.

There is some indirect evidence to show that the competent professional authorities warned the Coalition Cabinet of the dangerous course which they were pursuing, but the nature of these warnings was concealed from Parliament and the public, neither of which has adequate means or machinery for examining Ministers or calling for papers. Governments will die in the last ditch rather than authorise the creation of such machinery as the French possess in their Parliamentary Commissions for the examination of Estimates. These Commissions inspect Estimates, demand papers, call Ministers before them, and report to the Legislature. Few who have followed these reports over a series of years can deny that they make for efficiency as well as economy, nor yet can deny that Estimates are passed in our Parliament sometimes with slight examination, frequently with none at all.

Can we suppose for a moment that such Committees at Westminster would have passed without comment our Army and Air Estimates for 1922-23, or would not have suggested improvements in the plans announced ? Ministers can, of course, resign and give their reasons to Parliament. By not resigning they identify themselves with the policy of the Cabinet, and no subsequent excuses can relieve them. But professional members of the Board of Admiralty, the Army Council, and the Air Council do not, as such, resign if a plan is declared to be the policy of the Cabinet. It is then their duty to carry this policy into effect, however much they may differ from it and deplore it. Under such conditions no security whatever is derived by the public from the existence of the best professional advisers in high places. The Services are frequently laid on a financial bed of Procrustes and then amputated to fit the financial bedstead of the day.

If we wade through the Parliamentary Reports, a first point that strikes us is that we had no announcement of the relations between the Dominions and the Home Government on defence. We want to know what all parties are doing in this matter and we get no light. Before the war we were sure that the Dominions were working with us. The interchange of officers had good results. Our General Staff, under Lord Nicholson, had drawn close the links in Lord Haldane's day, and we knew that in organisation, armament, and equipment, not to speak of training, we had a common doctrine and a common practice. So when the Dominion forces joined ours in the war things went well. Now, we do not know exactly how we stand, except that in a political sense we cannot commit the Dominions in any way without consulting them. What are their present defensive arrangements, and are they satisfied with ours ?

We have also a greatly enlarged Colonial Empire of the old stamp, especially in Africa, and want to know what contribution it can make to the common stock in defence, and on what plan we are working. Again no light, and how many oversea territories are still under the Foreign Office, or under the joint rule of more than one Department, few can answer offhand. We ought to get the necessary replies under the Treasury vote, when defence can be investigated as a whole, but our past experiences do not make us feel confident that we ever shall. Our practice is to discuss the details first, namely, the Estimates, and the principles last on the Treasury vote. It is the cart before the horse.

If we turn to India, whose main element of defence is provided from home, we are astonished at the apparent indifference displayed by the Coalition Government to the situation on the Frontier, to the state of Central Asia, and to the internal ferment, tending to rebellion, which have all so greatly added to our military responsibilities. The last war with Afghanistan imposed upon us a greater strain than we have ever felt in India. Recent frontier expeditions have shown that the tribes are less easy to bring to book than they were. There is a lull in India internally, but our enemies have not disarmed. A recent All-India Conference 'reiterated its faith in the principle of non-violent non-co-operation for the enforcement of the rights of the people of India,' and this resolution was passed unanimously, all the members standing. It is in these circumstances that we diminish our Army in India, reduce our Expeditionary Force at home to a derisory standard, lower our reserves by 70 per cent. and take no steps to prepare the framework of a great national army against the moment, which may be to-morrow, when it will be needed.

Strange doings, indeed, and we have to add that our Navy is as blind in the air as it was during a large part of the war, and that the Air Force represents little but a training-school for its squadrons overseas, and has no pretence or power to combat any serious enemy in Europe, or to protect our shores from foreign raiders, or to repay such compliments in kind. Not once in the year 1921 did the Committee of Imperial Defence meet under the chairmanship of Mr. Lloyd George, and although the Standing Defence Sub-Committee met frequently, the value of its cogitations has not been apparent. The Navy, by the admission of Mr. Amery, has been cut down to the bone, the Air Force is incapable of playing a serious part in home defence, and the Army does not bear any comparison with 1914, when it was proved to be quite inadequate for our needs. So when we are told that 180 Cabinet and Imperial meetings took place during the year 1921 we are not much impressed.

It is difficult in such circumstances to feel a lively interest in the question whether a single Ministry of Defence should be substituted for our present arrangements. The first business of a

Government, advised by the Committee of Imperial Defence, is to provide the forces necessary to fulfil the objects for which they are maintained. We want clear thinking, and we want a plan of Imperial Defence, which is an indispensable preface to economical administration. No manipulation of departments and no discussions of sub-committees can reconcile us to the want of this plan and of these forces, nor atone for this want.

Commitments and Risks

In order to ascertain what character of armed forces we require in future we must first reckon up the new commitments which we have undertaken since the war ended and add them to those other commitments which are of prior date.

First comes our undertaking in Article X. of the Covenant of the League of Nations, 'to respect and preserve as against external aggression the territorial integrity and existing political independence of all members of the League.' This cannot be done by making speeches, and may imply the use of force. It is a large order, as there are some forty or more nations in the League. Our military policy amounts to a virtual renunciation of the intention to carry out the terms of the article, since we are no longer in a position to use force seriously for the purpose described and within acceptable limits of time.

Another new commitment is the Russo-German Treaty of Rapallo. We have guaranteed all the members of the League against aggression. Neither Russia nor Germany is in the League. If Poland, Rumania, or any other League State bordering on Russia and Germany is attacked we are bound to aid them. Russia, we have been told by Mr. Lloyd George, can mobilise 5,000,000 men. Germany has 7,000,000 veterans of the war. The concordance between these figures and our military provisions is not apparent.

It must be remarked that all the above Treaty commitments

are liable to become more formidable with time, when Germany and Russia revive, and when Austria, Bulgaria, Hungary, and Turkey see no hope of salvation, and of the recovery of their lost territories, except by joining the Russo-German compact. This compact has no time limit, nor has the League Convention.

A third commitment is the virtual independence which we have given to the Irish Free State. We hope for Irish friendship, but only time can show whether the hope is justified. By too many Irishmen England's danger is regarded still as Ireland's opportunity. We know very well by our experience of Ireland and Scotland in the sixteenth century, what it means to have a hostile people in these islands, a prey to the intrigue of every foreign enemy of ours, an instrument for vexing us, a footstool from which England can be overlooked. To fight another hard war with an unfriendly Ireland at our back, occupying a most potent strategic position in a maritime war, is not a proposition that we can contemplate with equanimity, but is one that cannot safely be eliminated from our calculations. The closure of South Ireland to recruiting for our armed forces is also an enfeeblement of our position.

We come next to the group of new commitments formed by the fresh territories which we have taken over from Germans or Turks, whether as owners or mandatories, in consequence of the war. These involve us, and certain of our Dominions, in important charges for civil administration and military protection, properly involving increased, and not diminished, military force. Our mandates in Palestine and Iraq have brought us into close contact with Arabs, Kurds, and Turks, and have made us peculiarly susceptible to the march of events throughout Arabia and in Asia Minor. Thereby the secure isolation of Egypt has been ended and the whole strategic situation in the Middle East has been changed. Those of us who opposed the campaigns in Iraq and Palestine naturally agreed with the late Mr. Bonar Law in wishing that we had never gone there, but cannot see clearly how we can evade the agreements into which we have entered respecting those territories.

Nor can we regard the future of Africa as perfectly assured. Political, religious, and racial troubles here find a fruitful soil, and already in North Africa we see the first symptoms of communistic propaganda among the natives. The white population in Africa is so small, and the native population so large, that Africa may bulk more largely in our eyes before many years have passed.

Europe itself is full of anxious problems. While the book of German reparations is not closed nor the security of France definitely settled, all the Continent remains under a cloud of fear of the consequences. The permanence of the German Republican Reich is still in doubt. The stability of Poland, Czecho-Slovakia, and Rumania is menaced by the Russo-German Entente. Europe is as stiff with armaments as ever she was. It is only the ownership of their armaments that has changed. Not one of the States of Europe has disarmed of its own free will. Over all hovers this menace of the Soviet rule, rendering all plans for limiting armaments futile until it has been laid. Through the Caucasus Moscow is in touch with Angora, and we have allowed the Turk to return to Europe. Greece reels under a financial and military effort which she undertook at Allied suggestion, but carried on without Allied support. Central Asia sways between Bolshevism and anti-Bolshevism. The Indian Frontier and contiguous States may at any moment require us to set large forces in motion. In the Far East China is honeycombed with internal sedition, and Japan has not yet regulated matters with the semi-Bolshevik Government of the Russians at Chita. Look where we will in Europe, Asia, and Africa, we see a multitude of difficulties and dangers with hardly one of which we can feel unconcerned owing to the immense extension of territories under the British flag.

A British Empire destitute of military force adequate to meet internal and frontier dangers, and without the machinery to recall a national army into the field, cannot hope to prevail in the political field nor to retain friends who look to the strong. We see the result to-day in the regional accords of the Little Entente, and in other accords between threatened States and France, which has by far the greatest and best-equipped army in Europe. We signed the Treaties in Paris and then incontinently disarmed, supposing apparently that others would do the same. This they have not done, except the vanquished in the war, and only they because compelled by force. Consequently our armaments, cut down to the bone in the case of the Navy, and through the bone for the Army and the Air Force, provoke neither the approval of friends nor the fear of enemies.

We were being led but recently and insensibly, perhaps by policy, perhaps by the force of circumstances, towards the gradual adoption of the American attitude of detachment from Continental Europe. Not even the great authority of Mr. Lloyd George in Conferences since 1919, nor his mastery in debate, induced Europe to follow his lead. But isolation, to be splendid, must also be strong. America is strong. We are not, though more exposed. A world of difference separates the two conceptions.

Isolation, however, is not a practical policy. Our interests are so intertwined with those of nearly all foreign Powers that we cannot disentangle them, and it is a fact that no Power, however powerful, can afford to stand apart and alone. Even if no wars, social revolution, or other catastrophes occur, we have before us a long period of patient endeavour to place the world in order, and we are as much bound in honour as we are in interest to play a great part in the reconstruction of Europe and the East.

Wars and Needs

The Conservative Party as a whole have not been distinguished in the past for attention to our armed forces, and our best modern Army reforms have been due to Cardwell and Haldane, two

Liberal administrators. The Coalition Government of 1918-22, by cutting down the Navy to the bone, by neglecting our Air Forces, and by carrying out reductions in the Army to the extent of rendering it almost useless for its tasks, had a still worse record.

The Conservative Party in 1923 had a chance of retrieving its reputation in matters of defence, but except the fixing of an adequate standard for our Home Defence Air Forces, it did very little, and for the Army it did nothing at all. The first thing necessary was that the Government should have had a clear and comprehensive understanding with the Dominions and India regarding the nature and amount of the armed forces of all Services which should be kept up in peace for the current needs of empire, or prepared for mobilisation in the event of serious contingencies. The more perfect the preparation for expansion the less may be the peace strength.

There are some who still cry out for a Defence Minister. If, by placing Navy, Army, and Air Force under a single Minister, we could make adequate preparation for war, there might be some sense in the proposal. But this mighty problem of Imperial Defence is altogether beyond the competence of the fighting Services, whether they are under one Minister or three. We have only to recall the number and importance of new Departments which the world war forced us to create to realise that a great war effects every branch of military and civil administration and all the public powers as well as every human being in the country. Instead of concentrating powers it is rather a question of decentralising them if the work is to be properly performed, and logical measures must be adopted to square with this consideration.

The Prime Minister is, and must always be, the true Minister of National Defence, in war as in peace. During peace, according to our existing arrangements, the Committee of Imperial Defence is his committee, and he can call to it such members as he pleases, many or few, official or unofficial. In war the practice of 1914-18 will assuredly be repeated, and he will become the head of a new

War Council. The exact composition of a future War Council will be for him to decide, but it will still be his committee, and he will preside over it. *Primus inter pares* he may still be called, but the fact will remain that he will be master, and that another Defence Minister will be superfluous, to say the least.

If we meditate over the kind of wars which we may be compelled to wage—and we shall certainly wage none except under great provocation, as we have far more territory than we need already—we may decide that they fall naturally into three categories. First, a small war within the Empire, partaking of the character of a military police operation, and capable of being undertaken by the Regular Army and its reserves and by the corresponding forces of India, and of such Dominions as decide to take a hand with us. Secondly, a more considerable war, like that in South Africa, which may require the use of forces larger than the Regulars can provide, but capable of being won by organised volunteering, without resort to compulsion. Third comes a great national war demanding an appeal to all the forces of the Empire, and entailing compulsion.

It is a great disadvantage if our Regular Forces are not capable of promptly and effectually dealing with internal commotions and frontier affairs, because delays, and still more the regrettable incident, are liable to cause trouble to spread, and the effort required, as well as the cost, to become much more considerable. We should, again, be able to place a respectable force in the field promptly, to meet our engagements under Article X. of the League Convention, and we have only to remember how many claims there have been since the war on the services of our Regulars in many parts of the world to understand that, as we happen to possess the greatest Empire in the world's history, we cannot afford to cut things so fine that the military police work of the Empire, or the prompt succouring of our friends, cannot be carried out.

The second category war, in which we have to appeal to

volunteering on a larger scale for recruits, is, of course, slower and more costly than the first, because most of the recruits have to be trained, and unless the enemy is only numerically formidable, this may take six months, except for such war veterans as may rejoin. The Territorial Army is naturally the first reinforcement after the Regulars have gone. If recruiting has to continue it is important that it should be organised, for a war of this character may develop into a great war, and we must not again enlist the key men of key industries, as we did in 1914 and 1915, because if we do we lose material which is indispensable at home for the prosecution of national war. Therefore this question must not be left out of sight, and recruiting must be organised on predetermined lines.

The last category of war, namely, national war, is too near to our memories for anyone to forget what it implies. We are the only Power that I can name, except lately-vanquished States, which has made no preparations for a national war, and has not even the vestige of organisation to meet the case. Even the United States has prepared, and I suppose that their risks of attack are not one-tenth of ours. We could even suggest that one in a hundred more accurately represents the comparison, and yet the United States has made this preparation and we have not. America has taken this line because she does not intend to be insulted and attacked again and find herself compelled to muddle along at huge cost for a year or two before she can deal with an aggressor, and this is the exact position that we are in from not following her example.

We have, it is true, the tradition of national service now, and enemies will no longer solace themselves with the thought that we shall never look at compulsion. But everyone abroad knows our armed strength to a battalion, and knows, also, that we have made no preparation for national war. This is a great source of weakness to our diplomacy, and is really a menace to peace and a cause of not a few of the regional and other military alliances contracted between nations. The kind of preparation for national war that we require is not costly, nor does it entail a call upon youth for service in peace. But it does entail very serious forethought and organisation, and the passing of certain laws like the old Registration Act, indispensable for the provident and timely collection of recruits, and it does imply the furbishing up of the Military Service Acts, which, when they have become law again, can be put aside for emergencies, as the old Militia Ballot Act was for so many years.

We want the world to know that we can send out half a dozen Regular divisions anywhere in a fortnight, and follow them up by fourteen Territorial divisions soon afterwards. We want the world to know that we have studied the manner of completing these divisions and maintaining their strength by voluntary service. We want to show the world that we have laid the basis for the organisation of a large group of divisions in case of national war, and shall resort to compulsion when the need arises. We are trading on an insufficient margin of military capital, and we have no reserve capital at short call. This is no fitting position for the Empire in a world still so heavily armed. Far from being an assurance of peace, it is a positive provocation of war.

It is necessary also that we should face the question of food supplies and not turn away our eyes from it. The late Mr. Bonar Law told us that we can have as much agriculture as we like if we pay for it. If the country wishes to be fed from its own resources in a national war, nothing will accomplish that purpose but protection or subsidies. If it will not look at these remedies for agricultural depression, then it must be told what the consequences will be in another great war, when our maritime routes can be assailed; and it must be prepared to provide the ships and aircraft to protect those routes and escort home the wheat which we do not grow in Great Britain in quantity sufficient for the food of the people. Since the Russian wheat supplies failed us, a single harvest failure in North America would produce disastrous consequences. No one has told our people that, nor has suggested a remedy.

There are also the questions of war material, and of raw materials generally. I do not recall that anyone has told us what war material we brought home from France, how much we disposed of, and how much remains. The figures have been given in America for the United States, but I have never seen them for our country. Many people are uneasy about our war reserves of material, and the figures should be called for. We should have a definite proportion for all the formations existing or to be raised, and when we see large sums voted in France for the reconstitution of war material we naturally desire to know our exact position, remembering how inclined our rulers are to let down stocks unless they are closely watched and are compelled to maintain them at a definite standard. If we are short of the raw materials needed by our factories in war, we need credits for their gradual collection.

Far from thinking that we should unite the fighting Services under one Minister in peace or war, I consider that our problem is to constitute a mobilisation centre in every public office which has the responsibility for supplying the Services or the civilian population in time of war, whatever the nature of these supplies may be. The thing is being done in France and it ought to be done here and be watched over by the Imperial Defence Committee. Our needs can be calculated from our experiences in the war and from the relation between the forces then mobilised and those which we propose to mobilise now. Because it is only an economic mobilisation it does not require less study than that of the fighting Services, and the right people to keep all the Departments up to the mark are the Committee of Imperial Defence.

Navy and Air

As we look over our defensive arrangements we find the Navy, cut down to the bone though it has been, in a better situation to carry out its mission than either the Army or the Air Force. The German navy is either destroyed or limited by Treaty in its future construction. No other navy in Europe rivals ours. Both France and Italy are financially unable at present, even if they had the desire, to build against us, except by the modern equivalent of the old *guerre de course.* With the greatest of our friendly rivals in distant seas we made terms at Washington, where a standard was set for us in capital ships.

It is true that we shall be glad when naval experts come to an agreement about the future of battleships, but so long as other Powers build such ships we must follow them and do the same. We cannot go very far wrong if we meet like with like in surface craft. We can go very far wrong indeed if we gamble in some new design and make a mistake. In other than capital ships and aircraft carriers Washington did not limit us in numbers, though it did so in displacements and calibres. On the whole, Washington afforded us a great protection against potential navy wreckers, while it enabled the United States to suspend the onerous competition in capital ships.

The one grave defeat in the Navy, in the opinion of most naval officers, is the absence of a Naval Air Service, under the control of the Admiralty, and both constructed and trained to co-operate closely and permanently with the Navy. The case for the Navy was admirably put by Rear-Admiral Sir W. R. Hall and Viscount Curzon in the House of Commons on March 16, 1922, and as admirably answered for the Coalition Government by Mr. Chamberlain the same day. It is indisputable that an air arm is necessary for the Navy in almost all conceivable future naval operations, and must be able to carry out every mission that the Navy sets it within the power of the air arm to execute. It is the same for Army work, and therefore both Navy and Army desire to have an air arm under their own control, and deplore the policy, for it is a policy, which prevents them from having it.

On the other hand, for the protection of this country against air raids, and for the delivery of counter-raids, the Air Force may have to play a virtually independent *rôle* when it may be better not to subject it to Navy or Army control. It was therefore thought best by the late Conservative Government, to meet all cases, that a trinity in unity should be established, and that the Air Force should be autonomous in administration and education, should be independent in resisting raids and in counter-raiding, and only be subordinate to naval and military commanders when naval or military operations are undertaken. But as all three of these war-like proceedings may occur simultaneously, it is possible that the decision may prove imperfect, even if we had the means for carrying out any one of them, which we have not.

If we cannot go farther than state the case in its general terms, it is because the question of naval and air co-operation has been under examination by the Standing Sub-Committee of the Committee of Imperial Defence and the full consequences of their decision are only partially known. But here again, the public will observe, it is, first of all, a question of whether we have the air arm we need, for, if we have not such arm, the duties allotted to it obviously cannot be performed.

The position in March, 1922,* when the Air Estimates were discussed, was that we had 31½ squadrons, of which 19½ were oversea in India, Iraq, Egypt, Palestine, and the Mediterranean, leaving 12 at home, only 3 of which could be devoted to home defence. We had 32 air squadrons, of which 15 were at home, including the units at Constantinople. The French had 128 squadrons, and the average establishment of our squadrons is 12 machines and of the French 9. The total number of aeroplanes of modern construction and fit for service has not been announced, nor have the respective programmes of the two countries been given to Parliament. This situation is most unsatisfactory. In our position nothing but equality with France will serve.

It is troublesome to all of us to have to quote the French figures, and we should be the last to credit hostilities with our trusty Ally. But we are here in presence of a question of prin-

* The latest figures are given in Appendix I.

ciple. In the old days we took for our standard of naval strength the two greatest naval Powers, and no one felt any ill-will towards us for doing so. Our insular position, our oceanic Empire, and our small volunteer Army placed us in a special position which all recognised. Our present naval standard is equality with any naval Power. But now that an enemy can attack us by way of the air as well as by sea, it is unavoidable that we should be equal at least in fighting aircraft to any Power within raiding distance * of England. No one fighting against us will bombard London if the immediate response is the destruction of his own capital. In the event that we evade the vexatious duty of rendering this possible we subject our policy to his, and are no longer masters of our own destinies.

There is scarcely a single military operation which is not profoundly affected by the presence of aircraft, whether our own or the enemy's. We need them for attack, for long-distance reconnaissance, for security, and for observation by day and night. It is difficult, demoralising, and costly to conduct operations against a foe who has a marked superiority in the air, and similarly his operations suffer the same inconveniences when the advantage is on our side. The air arm is peculiarly useful to us, who are so often obliged to act quickly in all parts of the world, for this arm is the most mobile that we possess. From the savage enemy, or from one poorly found in aircraft, our Air Force withdraws all hope of victory. For all these reasons, a large and highly efficient Air Force with our armies is a first necessity.

Many will agree with Mr. John Galsworthy that the best way out is to abolish all aircraft, as we have tried to abolish all submarines. That would suit us very well in Europe, though it would hamper us in colonial campaigns and savage warfare. But we should experience great and probably insuperable difficulties

* The Conservative Government accepted the principle of equality in 1923. It did not live long enough to get it, and the Socialist Government of 1924 rejected a motion embodying the principle, by the help of the Liberal Party.

with several Powers in reaching an agreement, and the only alternative is to show every nation that we do not intend to be out-distanced by any one of them. They might then be more likely to agree to prohibition, but we cannot consider that this likelihood is great. The conquest of the air by man is one of the great triumphs of modern science, and from such triumphs usually there is no going back. Air power must be regarded as a part of sea power, in the sense that the principle of equality with any single Power must apply. The French will not readily abandon the privileged position which they have gained by sacrifice and foresight. We must not expect them to do so. Their air force gives them an alternative means of bringing pressure on States which refuse to carry out the Treaty of Versailles, and we have no right to question their action. Commercial aircraft were, by a lapse of foresight, allowed to Germany under the Versailles Treaty, and may prove capable of adaptation to war and long-distance raids. There is no limit, under the Treaty, to such construction.

The submarine and the air arm have, in truth, played the deuce with sea power, and, failing abolition of both, we have to revise all our arrangements. Our sailors can be counted upon, with all the experience that war and research have given to them, to meet the submarine, given proper financial provision. So can the Air Council be counted on to raise and train any air force that the Government may sanction. The country is not going to lose its head, but equally is not going to sit quiet under a policy which places it, owing to the weakness of its air forces, in political subjection to any foreign Power, no matter how friendly that Power may be. It is useless to do this thing by halves. Abolition or equality are the alternatives, and there is no other. If the cost exceeds the amount which the Treasury can readily find, then economy in other directions must be sought to balance the account. The Navy, deprived of a new and indispensable arm, is no longer the sure shield that it was in the past, nor can home defence be counted upon without it.

The Army Needed

When our Air Force is permitted to bring up the air arm to an equality with that of any single Power within raiding distance of England, and when the Navy gives us a definite assurance that it is satisfied with the arrangements made for the co-operation of the Air Force with the Navy, we can turn to the Army and see what can be done to make it reasonably capable of meeting the heavy commitments referred to in this chapter. Without an Air Force that can hold its own and protect London, and without a Navy capable in all respects of mastering a hostile navy and of keeping open maritime communications for trade and transports, an army can do nothing. This is the foundation of all the strategy of an insular State and of an oceanic Empire. Therefore it must come first.

But these things do not win wars. An army wins wars. It always has and always will. And an army is still infantry with auxiliaries. Very important these auxiliaries are, but they remain auxiliaries. Artillery, cavalry, aircraft, tanks, machine guns, and all the other services are indispensable to armies, but it is infantry on which the greatest burden of the fighting falls. It is infantry that attacks the enemy, occupies his positions, thrusts him him back, and holds the conquered ground. Nothing compensates for want of a good, highly-trained infantry, and there is no equivalent for infantry but infantry. Obviously, the nature of the country where fighting takes place and the character of the enemy's army and defences regulate the nature and the amount of the auxiliaries. But in all armies the infantry division is the unit of measurement for military force. The auxiliaries, in due proportion, are assumed.

We cannot possibly accept without protest the preposterous standard of strength which the last War Secretary of the Coalition Government set for our Regular Army, nor the prehistorically dawdling mobilisation which he arranged for such Army as he left us. It responds to nothing real and nothing known. We must separate two things, the defence of the Empire against local attacks and rebellions, and national war, which is quite another thing. The British Regular Army has never been intended to provide for a national war. It is the military police of the Empire, which holds the vast territories other than those of our self-governing Dominions, garrisons the naval bases, and keeps a reserve at home to succour threatened points. Compulsion is wholly inapplicable to its functions, and after long and patient investigations the Cardwell system has been found that which is best fitted for its duties. If, under Lord Haldane, this reserve at home, constituting the old Expeditionary Force, was organised for war in a temperate climate and a civilised country, it was equally ready for service elsewhere.

How is it possible to regard our old six divisions and the two divisions of cavalry as anything but the irreducible minimum for the police reserve of the Empire in the known conditions of to-day? It is not possible. But vandals have destroyed this perfect weapon as an incidence of the campaign against squandermania, and not a Minister has had the courage to rise in his place and tell the public what it means. In hundreds of garrisons throughout the Empire, in every seat of local government overseas, in every place in dependencies and possessions where men meet, and in every General Staff of our Allies, the disillusion was profound.

The sixth division has gone. So has one of our two cavalry divisions. We cannot reconstitute these units when the battalions and batteries and squadrons have been disbanded. It is easy to destroy a Regular Army. It is not easy to build it up again with voluntary service, and the process is slow. Yes, certainly we must hold tenaciously to our ideas, and return to them when we can, but for the moment it is of more importance to insist that the four or five divisions left to us should be mobilisable in ten days. Whatever the General Staff require for this purpose should be accorded.

Similarly, we must reconstitute the Special Reserve system, whether we call it Militia or not, to supply a small reserve to the Army, and to incorporate, train, and despatch the reserves needed to keep our four or five divisions full up in the field. The Army Reserve must also be built up with all possible speed. That our present Regular Army will be found wholly inadequate to meet our needs on the first serious test, the writer firmly believes. For the Regular Army nothing short of the 1914 scale can suffice.

Worse, in some ways, is our failure to prepare the framework and lay down the organisation of a national army. Here the United States should be our model, and the American Defence Act of 1920 our guide. What we have to do is to recognise that in a great national crisis we must have compulsion as a basis, so that recruiting may be fair to all and may be equitably proportioned throughout the home territory. We have to lay down the plan, to allot officers and N.C.O.s to help in the first formations, to give the chance of training for commissions to civilians who are fit to lead men, and to store locally the arms and equipments for such formations as are destined to be raised in each locality. The Registration and Military Acts of the war time must be brought up to date, and the scheme must be presented to Parliament for discussion and then laid by for emergencies.

The Territorials can be adapted to play the same part in the system that the National Guard plays in that of the United States. Training must be dispensed with for a third line, as the cost would be too high, but opportunities must be given for it to those who wish for it. The national army may not be fit for service for six months, but as a great call on us is a probability of the future we cannot safely delay the preparation of the framework on large lines.

What we need is a great framework into which the civilian population can fit itself when a national emergency is declared by Parliament. We should prepare to restore all the famous divisions of the war time, sacred to our military history. Such an organisation, for fifty-four divisions, the American Staff is now preparing, with the approval of Congress, and a parallel organisation here would keep rebels and foreigners from fooling us.

Note on the Responsibility for the Army Reductions in the Year 1922

The Geddes Committee recommended a reduction of 20 millions in money, of 50,000 in establishments, and suggested Army Estimates of 55 millions. The War Secretary of the Coalition Government, namely Sir L. Worthington-Evans, reduced the money by £16,500,000, the establishment by 48,000, and produced Estimates of £62,300,000. In Vote A there was shown an establishment of 152,836 all ranks, compared with 201,127 in 1921. The Geddes proposals and the Army Estimates were not really comparable, because the latter included terminal charges for expenses due to the war, and also the cost of the new reductions, making a total of £10,500,000 ; and on the other side of the account they were relieved by the expenditure in the Middle East, no longer borne on Army Estimates, and by £1,500,000 of German money due for the cost of the Army of Occupation on the Rhine.

The Coalition War Secretary reduced the infantry by twenty-two battalions, including twelve from South Ireland, the cavalry by five regiments in addition to the four reduced in 1921, the artillery by forty-seven batteries, and other branches correspondingly. These reductions lowered the Army by 48,000 compared with 1921-22, and lowered the establishment by 20,000 compared with 1914. The Reserves of the Army, formerly about 200,000, including the Special Reserve, were down to 65,000. There was talk—it has for long been only talk—of the restoration of the Militia, but only £1,000,000 was taken in the Estimates for this purpose. So on the whole we were 155,000 Regulars and Reserves below the figure of 1914, assuming strengths to match establishments, and there was a great reduction in the Territorial Army.

What did the Coalition Government's policy for the Army amount to, after all? A reduction of a quarter of a million men below the 1914 figures, the hamstringing of the Haldane system, and the hopeless impoverishment of our fighting strength. The reduction of the fighting arms, and especially of the artillery, meant that we could only mobilise, at that time, four divisions in place of the six of 1914, and only one cavalry division instead of two. Instead of being able to mobilise six complete divisions and two cavalry divisions in a few days, we could only mobilise one infantry and one cavalry division in fifteen days, a second division within two months, and the third and fourth divisions in four months. The Coalition War Secretary admitted these facts. How many provinces and lives we should lose, and how many millions we should waste by thus dawdling into war is a fit subject for investigation by political actuaries. Shortly after this sacrifice of defence to finance was accomplished, the despatch of an expedition to Constantinople became the policy of the Government. Peace or war turned on the throw of a die. It was proved that the Coalition Government could not look six months ahead.

Moreover, the Special Reserve had been destroyed and the Militia not created. We cannot maintain units sent into the field for serious fighting purposes without the Special Reserve system of the Haldane plan or an equivalent. As for the Territorial Force, the War Secretary had the grace not to say when it could be mobilised. It may well keep the glamour of its fourteen imaginary divisions, though the fourteen mounted brigades have disappeared, but how can these divisions, low in establishments and low in strengths, and inferior in every possible respect to the standard of 1914, prove a force to be relied upon? Taking all these follies together, the Coalition Government can only be supposed to have been gambling on peace. To talk, as the War Secretary did, of 'having our finances destroyed and our powers of recuperation fatally injured' by refusing his reductions was very cheap rhetoric when it was a question, by his own admission, of only 16½ millions in the Chancellor of the Exchequer's Budget of £910,000,000. For all that followed from this policy the Coalition War Secretary, the Government to which he belonged, and the docile majority of March 22, must bear all responsibility hereafter.

The House of Commons contained in 1922, and contains now, a number of distinguished soldiers with recent experience of war in high positions. Not one had a word to say in favour of the Estimates, and all said, or implied, that they were appalled by them. The division list of March 22, 1922, is a paper to bear in mind. All those who voted for the late, but not lamented, Coalition Government share the responsibility of the latter for our military policy and for its future results.

CHAPTER XV

Churchillian Strategy

The Easterners—Mr. Churchill—Our Eastern Adventures—The Dardanelles—The Case of the Westerners—The Politico-Strategic Flank Attack—Wild-Cat Schemes—The Costly Errors of Our War Council, 1917-18—Conversion of the War Council by Defeat—The Conduct of the War in France—Inexcusable Action of Mr. Churchill in Preaching a Strategical Heresy at this Time of Day—Policy Made the Eastern Campaigns and the Policy was Wrong—When Policy Fails it Must Accept Full Responsibility for Failure.

THOSE of us who were resolutely opposed to the Easterners during the war, and attributed our great defeat in France in March, 1918, mainly to the political folk who had scattered our armed forces far and wide over the face of the earth, have kept silence since the Armistice in the belief that these gentlemen had recognised their folly. We now see by Mr. Churchill's second book on the war—namely, *The World Crisis*, 1915—that this is not the case, and there is some reason to suppose that we may have, from the same pen, dissertations on the same theme for the years 1916, 1917, and 1918.

Mr. Churchill is a man for whom many of us have a weakness. To him, and to Prince Louis of Battenberg, we thankfully ascribe the despatch of our Fleets to their war stations before the end of July, 1914. To him the Army owes much for rapid replacement of lost material in 1918, and again he undertook the arduous and thankless task of demobilisation, and carried it through. His courage, ability, and resource are points in his favour, while the absence of vindictiveness which characterises him contrasts favourably with the records of some other public men. But his judgment is usually at fault, and was never more so than when he became in 1915, as we see now, a leading advocate of the Eastern school, which recommended what he calls the 'politico-strategic flank attack' in preference to the settled convictions of all the greatest of the Allied soldiers.

This Eastern school affected with its poison very few soldiers, but it appealed to some of Mr. Churchill's political friends who had not made strategy a life study, and this deadly heresy led our armies to the Dardanelles, to Salonika, to Bagdad, and to Jerusalem, and to many other distant and unprofitable objectives. I suppose that we may, from first to last, have employed two million men or more upon these useless expeditions, with the result that the Army of France was starved for men, and met with the greatest defeat in its history.

This policy—for it was policy, and not strategy—was based on the illusion that there was a complete deadlock in France towards the close of 1914, and that we could afford to run about and try to win the war elsewhere without persisting in our endeavours to break the heart of our principal enemy. It was an illusion, because the German Higher Command never believed that the war could be won anywhere but in France, and returned to their original conception whenever the defeats of the Russians permitted it. Even after the German defeat on the Marne, von Falkenhayn endeavoured to accomplish the original plan by thrusting at Ypres. He failed owing to the want of training of some new formations which he employed there, and to the tenacity of the French, British, and Belgian troops. He came back again and tried once more at Verdun in 1916, and failed again. The Hindenburg-Ludendorf combination took up the same task in 1918 with a good superiority of force, and went near to a great success.

It was not till the summer of 1918 that the Allied and Associated Powers, by superhuman efforts, established the necessary superiority of force for victory. So far as our share in this superiority went, it was only due to the fact that our political chiefs saw at last that their Eastern policy was bankrupt, and that they hurried back all the troops they could collect from distant theatres, and sent every man whom they could raise at home to the Western Front.

Mr. Churchill's objective at the Dardanelles was a correct one. It was not reached because it was sought without settled plan, without proper preparations, naval or military, and because we did not at the time possess the trained military forces available and adequate for success. But the pusillanimous idea of winning the war without breaking the main German aggregation in France still attracted the politicians of the War Council, and the error persisted until the final result of the war—and it is the final result which alone counts—trembled in the balance.

The Case of the Westerners

We Westerners, from the first conversations with the French General Staff in 1906 up to the last day of the war, had a perfectly clear conception of what we wanted. We wanted to defeat the main forces of our principal enemy, persuaded that when this was accomplished his satellites would fall from him and the war would be won, whereas, in the contrary course of defeating the satellites, Germany would not surrender. We sought for, and readily found, the centre of gravity of these main forces of our principal enemy, and desired to concentrate against it every man and gun that we could place in the field until he cried for mercy, as he eventually did before the Armistice.

We desired to see a resolute offensive conducted on this decisive front until the principal enemy was beaten down, and to stand on the defensive elsewhere. We held in detestation all side-shows unless they were proved to offer quite extraordinary advantages, and we could only consent to them then if they could be conducted without prejudice to the main operations. The principle of concentration at the decisive point was our guiding star. We had in our minds the continuous development of new forces for employment in the decisive theatre until our superiority was overwhelming, and we only admitted the retention of strategic reserves in England according to the degree of uncertainty at the time, whether from the refusal of the Navy to guarantee us absolutely against raids or invasion, or from the natural caution which demands the retention of something in hand against unforeseen emergencies and the play of chance which is inseparable from war.

The course of the war brought the main German armies into a position where we could more conveniently deal with them than anywhere else in the world. The guns of the contending armies were heard on our coasts throughout the war whenever the wind had a slant of south in it. By rail, steamer, and barge we could bring our men and war material right up to our front with great rapidity. Our home bases were nearer to our armies than the German bases were to theirs. We could, and eventually did, make the Channel practically impassable to the German U-boat, and there was not that vast expenditure of tonnage upon transport and maintenance inseparable from distant oversea operations.

We could rapidly replace casualities and losses of war material, and quickly bring home our sick and wounded. We covered the Channel ports, and had a stroke been aimed at us at home, we could have rapidly transferred a corresponding force from France to England. We had, in fact, every reason in the world to congratulate ourselves upon the selection by the enemy of a decisive theatre so extraordinarily convenient to us.

That this conception of the strategy of the war was correct I for one have never doubted. It had, however, one immense defect : it was too simple for the politicians, and especially for that little knot of men who persuaded themselves and each other that they were great intellectuals and the soldiers narrow-minded dolts. Who first invented the seductive phrase about 'the politico-strategic flank attack'? Who was the genius that discovered the Central Powers were, in Mr. Churchill's words, 'vulnerable in an extreme degree on either flank'? Possibly it was Mr. Churchill himself. Possibly Sir Maurice Hankey, in his paper of December 28, 1914, on the Dardanelles. Possibly Mr. Lloyd George, in his paper of January 1, 1915, on a short-cut to victory over Austria. With much modesty these strategical prize essays are still withheld from us; and while Mr. Churchill labours the vulnerability of the enemy's flanks, he very carefully and astutely abstains from giving any account of any plan through which we could profit from this vulnerability, while his own attempt at the Dardanelles was an abject and costly failure.

It is of course necessary for his ease, before producing his quack remedies, that he should by every artifice, adjective, and expletive seek first to produce in the public mind a feeling of contempt for the soldier's strategy. So we get his picture of 'the swaying and the staggering,' the 'helpless violence,' the 'slaughtering and the squandering,' the 'sanguinary war of exhaustion,' the 'dull carnage,' and much other unmitigated balderdash, of which we can only say that it is never necessary to indulge in this type of fudge save when an advocate has a thoroughly bad case. I have not yet discovered how Mr. Churchill expected to work through the 13 or 14 million Germans who took part in the war except by killing them. He would not attack the Germans, not he! He plumes himself over the Dardanelles in saying that 'neither by sea nor land was the same formidable German resistance to be expected.' But really, when 13 million Germans want to fight us and our Allies there is no advantage in going where they are not to be found. Nothing aroused deeper satisfaction at German Headquarters during the war than the dispersion of our resources in the East.

The Politico-Strategic Flank Attack

These wild-cat schemes of politico-strategic flank attacks were the offspring of political dilettanti ignorant of the whole nature and meaning of war and of its needs. Napoleon never attempted them. Strategical success is the successful preparation of tactical victory, and we still await Mr. Churchill's explanation of how the Central Powers were to be beaten by his methods. To declare that frontal attacks have been abandoned for forty years, and to contrast them with the success of the Japanese turning movements at Liauyang and Mukden, displays either ignorance or a desire to deceive. At Liauyang the Japanese armies made desperate and long-continued frontal attacks, and at Mukden four armies out of five assailed the Russian front from February 19, 1905, till March 1, when Nogi's flanking movement began. These frontal attacks permitted, and were the necessary preliminary to, the turning movements.

The only episode in the World War at all resembling what Mr. Churchill dimly hankers after was the entry of Rumania into the field and the speedy advance of her armies into Transylvania. That was met, when they came within reach of the railway zone of the Central Powers, by a prompt concentration of enemy forces, and had we depleted France at any time it was open to and legitimate for Germany to despatch a corresponding force to oppose us. To suppose that a strategic flank attack necessarily leads to a tactical flank attack is a delusion of amateurs. If the use of the sea for transport was an advantage until the U-boat made it too expensive, it is also true that the politico-strategic flank attack, if it were to effect anything, must in time have had to deal with the railway network of the Central Powers, which was very formidable indeed.

These transmarine operations on a large scale proved fearfully costly and much open to U-boat attacks. It needed 740,000 tons of shipping to convey ten divisions to Salonika. We found that it required four tons a man all told to carry a force, and the permanent employment of three tons a man afterwards for maintenance. The larger the number of ships the greater the toll exacted by the U-boats. Far from our bases in distant seas, we could not rapidly replace casualties, as Sir Ian Hamilton had good cause to know, and we were bound to have landed in countries where the language, communications, and local supply created enormous difficulties.

We did not find in the East the vulnerability of the Central Powers, and we were reduced at Gallipoli and Salonika to those frontal attacks which are the special objects of Mr. Churchill's scorn when they are attempted on the Western front. We never succeeded in subduing the German will by any of our distant campaigns, no matter what renown they brought to the generals who won victories for us.

Mr. Churchill endeavours to show that the Dardanelles adventure was launched in order to aid the Russians, who were alarmed about their Caucasian front. That sounds a very reasonable, or at least plausible, defence until we examine the question. Then we remember that the superiority of the Russians over the Turks on this front was re-established by the battles fought at Sarikamish and Ardahan, between December 25, 1914, and February 4, 1915. These battles completely checked and threw back the Turkish attacks, and the result was known a fortnight at least before the bombardment of the outer forts at the Dardanelles began. The absence of all allusion to these facts by Mr. Churchill is a trick played upon the public memory.

It is, on the other hand, probable that had not all our distant campaigns been undertaken, we should have been strong enough to prevent the Germans from detaching such considerable forces from the Western front to destroy the Russian armies. The whole German plan of campaign had been ruined by the victories of the Marne and Ypres.

I have a German return which shows that the German losses up to the end of March, 1915, had been 1,214,523 killed, wounded, and missing. It was partly the alarm created by the severity of these losses that made the Germans seek for softer material on the

Eastern front, and adopt a defensive attitude in the West. In fairness to Russia it was our duty to continue and gradually increase the pressure in the West, and not to direct our armies to distant theatres where they affected German strategy not one whit. The German line in the West was 468 miles in length, and it was true in the past, as it was in 1918 when Ludendorff smashed our 5th Army, that when a line is unduly extended it can be penetrated. When the Americans came along they at once displayed a healthy and rooted contempt for the side-show and all its works. They went for the gloves, sought the centre of gravity of the principal enemy, found it, and by attracting to themselves 47 German divisions between September 26 and November 11, 1918, contributed greatly to the final victory.

Things having passed in this manner, and our defeat in France having proved the absurdity of the vague tenets of the Eastern school, we are entitled to reject with derision Mr. Churchill's claim that they were 'vindicated in the end.' They were never vindicated. They were political campaigns undertaken in complete ignorance of the rudiments of strategy, under an entire misapprehension of the situation in France, and in direct opposition to the best military advice. They were associated with the wicked depletion of our forces in France, and they cost us hundreds of thousands of casualities and much of the debt under which we now groan.

The spokesman of the War Council declared on March 7, 1918, that 'there will be no dangerous superiority on the Western front from the point of view of guns any more than from the point of view of men,' and stated that he was 'still a little sceptical' about the threatened German offensive, which was launched a fortnight later. By what bad advice Mr. Bonar Law had been bemused into making such a statement to the House of Commons we still know not, but we know what followed. After the Versailles Conference of June 1 and 2, Mr. Lloyd George and the other Allied Prime Ministers sent a message to the President of the United States, from which I take the following passages:

'The crisis, however, still continues. General Foch has presented to us a statement of the utmost gravity, which points out the numerical superiority of the enemy in France, where 162 Allied Divisions now oppose 200 German Divisions, is very heavy, and that . . . there is great danger of the war being lost. . . .' That was, in fact, the result to which our Eastern fatuities had all along steadily pointed, and had led up to from 1915 forward.

The Conduct of the War in France

I do not know why Mr. Churchill should give such hard names to the wearing-out battles in France and to the war of attrition which all our armed forces conducted. A war of exhaustion is not a novelty. It was preached by Clausewitz after the Napoleonic wars as the best means for increasing the waste of the enemy's force.

'The idea of wearing out in a struggle,' he says, 'amounts in reality to a gradual exhaustion of the physical powers and of the will produced through the long continuance of exertion.'

We could not swallow at a gulp the most colossal and formidable military organisation ever placed in the field, and the process followed on the Western front, and by our Navy in its sphere, was perfectly in consonance with the requirements of the situation. It would have been done better and more quickly had not our Eastern school diverted such large forces to distant theatres.

We are not called upon to defend the direction given to every attack during the long struggle. The situation in some respects only unfolded itself gradually, and everyone had to learn his lessons. Personally, having been Military Attaché in Belgium and Holland, I was always dead against the attacks made in Flanders, not only on account of the ground and the water, but because no success on this flank promised great strategic consequences. Also, until our superiority in men and guns was very considerable, I was convinced that we should restrict ourselves to limited offensives.

From a strategical point of view, the correct direction for a great effort seemed to me obviously on the Meuse-Argonne front, which General Pershing eventually selected. The Carignan-Sédan-Mézieres railway was indispensable to the Germans for the transport and supply of their troops, and could a force place itself on this line and maintain itself there before the Germans could withdraw their forces, their ruin would be consummated.

But these things represent personal preferences, and the great aim of the Allied Armies in France was to wear the enemy down, which process was steadily pursued from the first day of the war to the last. Had our armies been kept up by our War Council, and had this body cut itself adrift from the Churchillian Eastern diversions, it is possible that the final success might have been obtained a year earlier. What is, however, perfectly inexcusable is that Mr. Churchill should not realise, even to-day, that he is preaching a rank strategical heresy which was completely exploded by bitter experience, and that he should employ his undoubted talents in leading the English people astray.

Not one of the four great adventures in the East was approved by the General Staff. Policy made these compaigns, and the policy was wrong. Too well, alas ! we know that policy governs strategy, but when policy fails it must not attempt to foist its failures upon the fighting Services, but must accept the responsibility which it has incurred, and stand up in the dock for sentence.

CHAPTER XVI

Singapore or Sydney?

Genesis of the Plan for Constructing a Naval Base for Our Grand Fleet at Singapore—The Dominions and India in Agreement—No Justification of the Plan yet Given—The Capital Ship—The Washington Conference—Singapore—Must Remain a Cruiser Base—Inadequate Local Facilities at Singapore for a Grand Fleet Base—Strategical Objections—Extensive Defences Needed There—No Signs of Them—Japan's Gross Tonnage—A Serious Enemy—The Port Arthur Precedent—Great Distance from Sydney—The Old Strait at Singapore—An Unsuitable Site—Government Plans—A Division of the Fleet—Previous Assurances that it Would Not be Divided—We are Following the Fatal Russian Example of 1904-05—Sydney—The Spacious Harbour of Port Jackson—Facilities of Every Kind—A British People at its Back—A Central Position—Concluding Remarks—The Main Naval Object of Japan Before Contact Between the Main Fleets—We Shall Not Fight a Great War with Fleets Alone—The Armies of the Empire will Concentrate in Australia—The Grand Fleet at Sydney will Cover this Concentration.

THE genesis of the plan for the construction of a great naval base for our Grand Fleet at Singapore has not yet been completely revealed to us, but the plan apparently first took definite shape in the year 1921, when Mr. Lloyd George was Prime Minister, Lord Lee First Lord, and Mr. Amery Financial Secretary to the Admiralty. We must assume that the plan originated with the Naval Staff. The Foreign Office has denied that it made any representations to the Admiralty that a new naval base was required at Singapore, but naturally the Foreign Secretary is responsible, like his colleagues, for a Cabinet decision.

The plan was submitted later in 1921 to the Imperial Conference held in London in that year, after it had received the approval of the Cabinet and of the Imperial Defence Committee. The Conference is said to have accepted it, but the official report upon that Conference (Cmd. 1474) has no reference to the subject. What probably happened was that it was considered advisable to await the result of the Washington Conference before making an announcement. The plan again came before the Conservative Cabinet when the Coalition Government had disappeared. It was again accepted. As all the Dominion Governments and India had apparently been in agreement, and as the Board of Admiralty was strongly constituted, there was every reason to suppose, as indeed has happened, that Parliament would accept it.*

If there is opposition, it is mainly because the naval authorities have not yet produced any justification for the plan, whether by giving strategical reasons for the selection of Singapore or by showing that the local conditons are favourable for the project. These are the real objections, and they have been apparently reinforced, but in effect weakened, by the campaign against battleships, led by Admiral Sir Percy Scott and others, and by the contention that the plan is against the spirit of the Washington Conference. Singapore has in fact, and to the great disadvantage of the public service, become a party question, and promises to become a hardy annual of politics, since moneys will have to be voted from time to time as the work goes on. The total cost of the scheme is given at ten millions, but only comparatively unimportant sums are to be spent during the first two years, and the cost of defences, garrison, and future maintenance has been ignored,

The Capital Ship

It is not worth while to go deeply into the question of capital ships *versus* submarines and aircraft. So long as the other leading Naval Powers, and especially Japan and the United States, hold to the capital ship and make it the main item in their programme, it would be a great risk for us to strike out a new and more or less untried line. The capital ship, in the opinion of the great majority of the best seamen in the world, was the foundation of our victory

* The Labour Government of 1924 abandoned the plan.

in 1914-18, and so long as that view holds the field in the greatest navies we should be gambling with safety to replace it with some other instruments and means for conducting naval war. It is partly on account of the newer bulged battleships that we need fresh graving-docks in the Pacific, and it is clear that to send these ships to the Pacific and to be unable to dock, clean, and repair them would be a most foolish proceeding.*

It may be true that the submarine and aircraft may overcome the capital ship in the future, or that some other development in naval attack and defence may make prospective docks as useless as the old ones have become for the bulged ship. But it has not yet been proved to the satisfaction of leading sailors that submarines, and still less aircraft, have sounded the knell of the capital ship. It is true that the Germans devoted their submarine activities mainly to the modern form of the old *guerre de course*, and not specially to the assault upon warships, or what Admiral Scheer calls military action.

Our next enemy may vary this procedure, when the results may also vary ; but, all things considered, sailors hold that there is no good or sufficient reason to condemn the capital ship on our experiences in the last war, and therefore it is right and necessary that we should construct a properly equipped base for them as near as we can to any potential theatre of naval war where we have great interests to defend. The public certainly will support the Admiralty when it demands such base, but with the reservation that the strategic necessity is made out, and that the locality selected is suitable for the purpose from all points of view.

The Washington Conference

Much has been made in Parliament of the Washington Conference and its supposed implications. There is no proof that a

* The Naval Architects at their Annual Meeting in 1924 were told by Sir Philip Watts that there was no advantage in giving bulges to a new ship, and it was said that the new battleships would not be fitted with them.

great naval base at Singapore, whatever we may think of it in other respects, violates the spirit of the Treaty signed at Washington on the limitation of naval armaments. The text of Article XIX. of that Treaty is perfectly clear. It was obvious to everyone at Washington, where I happened to be during the Conference, that the choice of the meridian of 110° east longitude as the western limit of the zone within which—with certain exceptions specified in the text—no new fortifications or naval bases were to be established, was deliberately agreed upon to leave us complete liberty at Singapore.

We and the United States had given up all idea of improving Hong-Kong, Manila, and Guam, and had thereby relieved Japan of all fear that our policy aimed at the preparation of great naval bases within comparatively short striking distance of that country. By withdrawing to Singapore and Hawaii, which are respectively 3,000 and 3,440 miles from Japan, we gave abundant proofs of our pacific policy. But, as the United States took a compensation by excepting Hawaii Islands from the restrictions of Article XIX., so we compensated ourselves by excluding Singapore by virtue of its position in relation to the meridian of 110° east longitude.

If our Delegation had entertained the slightest suspicion of the political capital that would be made in London out of this arrangement, they would, no doubt, have taken the meridian of 100° east longitude, and then have excepted Singapore as the United States had excepted the Hawaiian Islands. But nobody at Washington suspected for a moment that such a really preposterous inference would be drawn from a clear and straightforward arrangement.

Singapore

But when we come to inquire why the Admiralty fixed upon Singapore as the base for the Grand Fleet in the Pacific, we can get no reasoned answer. It is quite true that Singapore is one of the most interesting ports in the world, and that in normal times its trade, mainly a transit trade, is very great, partly because it is a free port and exacts neither harbour nor tonnage dues. Being such a great trade centre, and in a very favoured position, it has naturally become a naval and military station, and has been for many years. There is no question that, in the event of a war in the Pacific, Singapore must remain a base for cruisers, destroyers, submarines, and aircraft, and should be more strongly defended and garrisoned than it is now.

But we must bear in mind that in the event of war with Japan, all the Singapore trade, with Japan first, and then probably with China, would be paralysed, and that much of the trade of other countries in the Pacific would pass west by other routes. To talk of Singapore as the gateway of the Pacific, as though there were no other, would not be true in war, as we surely learnt during the last war, when von Spee's cruisers left the Pacific by a different route. The corrective to an undue appreciation of Singapore in the event of hostilities is a study of a chart of the Pacific and of its innumerable ocean routes.

When we examine the choice of Singapore as a great base for our Grand Fleet in war we are astonished at the selection. The Grand Fleet is an immense and imposing organisation, as those who saw it in the war time will recall, and it needs for the exercise of all its powers a wide area of well-protected water space which Singapore does not provide. It needs security, food, an industrialised population, and a score of other things which Singapore does not supply; while the climate is hot, damp, exceedingly enervating for Europeans, and specially for young men.

Singapore is indeed a gateway in the sense of being on the outer edge of the Pacific, and not in it. Its position is eccentric in relation to our greatest interests in that ocean—namely, the defence of Australia and New Zealand, not to speak of the Pacific shores of Canada. It is 8,000 miles from England by the Mediterranean and 12,000 by the Cape route. It is 3,000 from Japan. If our Fleet left England *viâ* the Mediterranean at the same time that Japan's Fleet left Japan, and steamed at the same speed, our Fleet would be 5,000 miles distant when the Japanese Fleet reached Singapore. Either one Fleet or the other might make an earlier start or be delayed, but in any case Singapore must be able to hold out for a considerable time against a serious attack, or it may fall into Japanese hands before our Fleet arrives.

Defence for some considerable time can indeed be expected if adequate measures are taken, but as yet there are no signs of them. Without them, the chances are that Singapore will fall, and the Straits of Malacca and the Sunda Strait will swarm with Japanese submarines and mines before our Fleet arrives.

We must remember that Japan is three times stronger than when she beat Russia single-handed in 1904-05, and that she has made preparations for great expeditions oversea, and has 3,000,000 gross tons of mercantile steam shipping at her disposal. In the interval between the outbreak of war and the arrival of our Fleet in the Pacific, Japan will naturally take such measures as she sees fit to improve her position. What those measures will be I shall not attempt to investigate, but many alternatives will be open to her, and we can be sure that the choice will be as disagreeable as possible for us. Let no one suppose that our friend Japan would not be a very serious enemy. She has a tradition for striking at, if not before, a declaration of war, and has proved capable of capturing what the Russians called 'the most impregnable of all first-class fortresses'—namely, Port Arthur—in seven months, though the fortress, with its 59 permanent forts, was defended by some 45,000 men of the land and sea forces with 546 guns.

The chief bait for the Japanese in that case was the remains of the Pacific squadron in the harbour of Port Arthur. In the case of Singapore, it would be our new docks when they are made, and any detachments of our Fleet that might be sheltered in that confined ditch, the Old Strait, where they would be the passive mark for all the Japanese aircraft.

Even if, by some lucky chance, our Grand Fleet had reached Singapore before an outbreak of war, it might hear that Melbourne, or Sydney, or Wellington, was attacked, and then might have to steam another 4,400 miles to reach Sydney, passing—if it took the shortest route—through a maze of islands for 2,560 miles before it reached Thursday Island and traversed the Torres Straits into open water. A better opportunity for submarine attack could not be desired. Our Dominions in the Pacific are not, in my humble judgment, defended by our Grand Fleet at Singapore, but if the Governments of these Dominions think they are, then theirs will be the responsibility for the opinion and for the future consequences thereof.*

From the point of view of local facilities for a great naval base, Singapore has few claims. The Admiralty on June 20 1923 gave out that the Old Strait had been selected for the site of the new dockyard, and the Straits Settlements, with their customary generosity, made us a present of the site. The Old Strait separates the island of Singapore from the Malay Peninsula. The western approaches to it are barred by sandbanks with depths of three fathoms, while the fairway within has a depth of four fathoms in places.

The eastern entrance has better depths, but the island side of the Strait has no facilities whatsoever for a base, and everything connected with such an organisation must be created. Incidentally, the choice made leaves the new base outside the existing defences, and exposes it to attack from the side of the mainland if a hostile force lands there. The only thing that can be said of

* Our readers may recall that in May and June, 1910, Colonel Repington wrote two articles in *Blackwood* on 'New Wars for Old,' foretelling the great future of submarines and aircraft in the next war, a forecast abundantly justified in 1914-18. He also named Scapa Flow as the proper base for our Fleet in a war with Germany, and advocated the use of 'that poaching expedient the otter' for service against submarines and mines He was the first, so far as we know, to advocate this contrivance, which proved of great utility in the war.—Ed., *Maga*.

this site is that it is just about the most unsuitable one that can be imagined. It is a paradise only for contractors.

Government Plans

I do not think that any great harm will be done by the modest programme of construction which Mr. Amery has announced for the immediate future. A floating-dock, a supply basin where lighters and transports can unload their cargoes, repair ships, together with armament and munition stores, will do very well for a cruiser base, and for submarines, destroyers, and aircraft, which are the arms indicated for use in these narrow waters. That must come in any case, so the Government have time to look round and revise their views in the light of experience and local knowledge.

We are promised 'a by no means powerful force of cruisers,' to which also there can be no objection; but when the First Lord speaks of two or three capital ships which are also to go out, the case changes. We are then travelling far away from the sound position laid down by the Parliamentary Secretary to the Admiralty, Commander Eyres-Monsell, in the House of Commons on May 1. Then he told us that 'we cannot divide up the Fleet as we used to do in pre-war days. The Fleet must be kept intact.' We are now, on the contrary, dividing up the Fleet, the main portion destined for the Mediterranean, part at home and on foreign stations, and lastly, the detachment in the Pacific, which may include three of our now limited capital ships. We are somewhat unpleasantly reminded of the Russian naval strategy before the war with Japan.

Sydney

All this cost and turmoil respecting Singapore need not have been, to the great relief of the Government and the avoidance of international friction, had the last Coalition Government gone to the Australians and invited them to build quietly a few dry docks for bulged capital ships in some corner of Sydney's harbour of Port Jackson. This great harbour, one of the finest in the world, already possesses in abundance all those things, except bulged-ship docks, that we propose to construct at great cost at Singapore. Wharves and quays, hospitals and warehouses, factories and foundries, all exist in plenty, and there is every facility for repairing vessels of all kinds, both stores and supplies being abundant.

It is a vast harbour, extremely defensible and secure, with excellent anchorages, and it is above all an ocean port very inconvenient to attack for hostile submaries. The Grand Fleet can get there by ocean routes little liable to surprise, and on reaching Sydney it will find a British population at its back, and all the resources and energy of Australia behind it.

It may quite fairly be objected that Sydney is 12,000 miles from England by the Cape route. So it is, and 4,420 miles from Japan. Therefore if the Japanese make for Sydney they can get there while our Fleet is still some 7,500 miles away. But whereas at Singapore we have no British to draw upon for defence, at Sydney we have the whole population of Australia, which has proved in war to be equal to the best elsewhere, and still has 300,000 veterans of the war, besides the younger contingents trained by national service. With these Sydney can assuredly be made safe for the necessary three months. So the extra distance does not matter, and besides, if our Fleet is called to Sydney from Singapore, it will equally have to cover over 12,000 miles from England before it gets there.

I believe in making Australia initially the Empire naval and military base in case of a war in the Pacific. We cannot safely make it so with our main Fleet 4,400 miles distant up in a dead angle of this ocean.

I fancy that some of the objections which come to us from Australia respecting my proposal of Sydney arise from a fear that the trade of Sydney will be interfered with if the Grand Fleet arrives there. A war in the Pacific with Japan is not likely to leave Australian trade immune, but such trade as there is with China and Japan will cease in any case, and I am disposed to believe that we can get on very well at Port Jackson without interfering with the fairway or with the Sydney trade centres at Darling Harbour and the two other coves which handle Sydney's maritime business.

Sydney occupies a central position in relation to our main interests in the Pacific—namely, the defence of Australia and New Zealand—and has as many local advantages as Singapore has the reverse. It therefore seems the most natural and fitting war-station for our Grand Fleet in the initial stage of a war with Japan, and if we trace out the subsequent operations we may find it no less suitable in all contingencies that can be foreseen. Why then should the Admiralty prefer Singapore, and the Dominions and our Cabinet back them up?

Concluding Remarks

I think, first, that the case for a great naval base in the Pacific is made out, and, if that be admitted, then the Cabinet—any Cabinet—would be naturally disposed to leave the choice to the sailors. Singapore is an attractive spot, and when people arrive there after a long voyage there is something alluring about a busy port swarming with shipping of every kind. Singapore seems to our fascinated eyes the Metropolis of the East. But, as I have ventured to suggest, that busy trade will disappear in war time, so far as Japanese and Chinese trade is concerned, if Japan takes the measures which I imagine she would take; and then, without its maritime trade, Singapore becomes a secondary interest.

The Dominion representatives are usually disposed to fall in with any naval arrangements suggested by the Admiralty, which is a very authoritative body for them. Singapore as a naval panacea has also this particular attraction for Anzac statesmen—namely, that it does not burden their budget, but is a charge on our taxpayers at home. If our sailors tell them that the Grand Fleet at Singapore covers Australia and New Zealand by virtue of some flanking virtue, then the responsibility of the Admiralty is engaged ; but if Dominion representatives take this opinion without using their brains, and there is hereafter found a fallacy in the claim, then the responsibility towards the Dominions rests with their own representatives.

I am not ignoring the fact that New Zealand proposes to subscribe £100,000 for Singapore, but also for defences in New Zealand. Australia may do the same, but this will be but a drop in the ocean of the eventual cost. What this cost will be no one can tell. We know the estimate, of course, but we do not know how many dry docks it includes for the bulged ships. On the Tilbury new docks the Port of London Authority proposes to lay out £5,000,000, and their plan includes, in addition to a large wet dock, only one dry dock, of apparently the same length, breadth, and depth over sills as a bulged ship needs.

How many similar docks are needed for the Grand Fleet in war, when sixteen of our existing capital ships already need them ? We are not told, but obviously we are in for a fairly heavy expense on the graving-docks alone. In the last war two of our battleships were always refitting, apart from damage in action.

As for the famous flanking position and the idea that we can, with any certainty, intercept a Japanese armament upon its way home after an attack upon Australia or New Zealand, I think that the chances of interception in such vast expanses of ocean are not great. We remember Scarborough and Hartlepool in December, 1914, and how, with all the necessary information at our disposal in advance and with our ships suitably placed between von Hipper's battle-cruisers and their home base, the act of interception failed owing to poor visibility. If we are told that the Pacific is not the North Sea, we can certainly agree ; but, with regard to visibility, on the day of Togo's victory over the Russians in the Sea of Japan, 'a heavy fog covered the sea, making it impossible to observe anything at a distance of over five miles,' as Admiral Togo reported. It was mainly because the Russians sought battle and steamed towards narrow straits where the interception could not fail to take place that battle was joined.

I do not like this idea of trailing out our capital ships to the East through a succession of straits and narrow waters to a base in other waters also admirably suited for the action of submarines. The Straits of Gibraltar, the Mediterranean, the Suez Canal, the Red Sea, the Colombo focal point, and the Straits of Malacca, are all points where a well-advised enemy would prepare his mines and locate submarines to try to wear away our strength before battle. This will be the main naval object of Japan before contact between the main Fleets. Let us get back to the ocean and recognise that Sydney is an ocean port, and can be reached by various ocean routes and not the Cape route alone. I do not wonder that Admiral Tosu, the Japanese Naval Attaché in London, has given out that no better locality for a dockyard could be chosen than Singapore. From the point of view of Japan he is absolutely right.

Last of all, our people must reflect that we shall not fight a great war in the Pacific with a Fleet alone. The armies of the Empire will concentrate somewhere in that ocean to oppose the enemy, and where should they concentrate but in Australia? This act of concentration may take a long time before our armies are all assembled and prepared. Singapore may be a suitable point of concentration for India's contingents because the climate is less unsuitable for them, and the restricted area convenient for moderate-sized forces; but the British and allied forces will inevitably go to Australia until the direction of the military effort is decided.

The Grand Fleet at Sydney will cover the concentration and defend the avenues of approach, whereas at Singapore it will be almost useless for these purposes. To Australia will also return from the front the sick and wounded to be tended, while all the vast impedimenta of great armies can there find suitable localities for a base. When, later, the armies are sent on to the points of attack, they can be so sent in secret from Australia, under protection of the Fleet, given proper precautions in advance, whereas every movement from Singapore is liable to be reported by the Chinese and Japanese spies who will swarm there. Thus from all points of view Sydney should be the Pacific base for our Grand Fleet in time of war.

CHAPTER XVII

The N.W. Frontier of India

India's Neighbours

I.—1800-1810

A Record of Our Relations with Afghanistan and Persia, 1800-1923—Napoleon and Persia—Gardane's Mission—Tilsit and Persia—The First Afghan War—Dost Mahomed and Persia—The Second Afghan War—Abdurrahman—The Curzon Policy—The Latest Phase—The Lawrence Policy Holds the Field.

In order to estimate at their true value the present relations between the British Government and the Government of India on the one side and Afghanistan and Persia on the other, both Parliament and the public are in need of a Command Paper which will describe, fully and faithfully, the negotiations which have taken place since 1919, when the Amir Amunulla broke away from our control of his foreign relations and opened a new chapter in Central Asian politics.

But, if we are dependent upon the Government of the day for this information, we are not so dependent upon it for the history of the past, and it is indispensable to keep before us a summary, however brief, of the main events of this history in order to comprehend the present and to avoid some of the errors and miscalculations which have marked our relations with Afghanistan and Persia since the opening of the nineteenth century. It was only in 1800 that these two States first began to arouse our interest in India and to attract the serious attention of our diplomacy.

In 1800 Wellesley was Governor-General, and our political frontier in India was far away from the mountains, and even the Indus. Marengo was fought in that year, and Bonaparte's star was in the ascendant. Neither Bonaparte, nor we ourselves, knew much about Persia or Afghanistan in those days. The Durani Empire consisted of the present Afghanistan, part of Khorassan, Kashmir, and the Derajat, and over it ruled Zeman Shah, grandson of Ahmed Shah. His power was over-estimated because of our ignorance of his resources, but by posing as the champion of Islam he appealed to the Mohammedan princes of India, and as his reign synchonised with the rise of Bonaparte and the Napoleonic wars which led to French activity in Persia, the general situation caused some anxiety. Zeman Shah had planned a great invasion of Hindostan, but he did not penetrate farther than to Lahore. A revolt called him home, and in 1802 his brother, Shah Shuja, was on the throne.

The Tsar Paul and Napoleon were set upon a great Franco-Russian invasion of India. Paul's plan was for 35,000 Russians and Cossacks to concentrate at Astrakhan, while 35,000 French under Masséna were to be detached from the army of the Rhine and to descend the Danube, and then cross the Black Sea and the Sea of Azov. The combined forces were then to cross the Caspian, to traverse Persia and Afghanistan, and to emulate the exploits of Nadir Shah. That the plan was seriously examined by Bonaparte seems certain, but Paul's death brought the great project to the ground in 1801. Napoleon, as Emperor, continued to toy with it from 1803 to 1810. He sent General Decaen to the Indian Ocean, where this general did us all the harm he could until the Ile de France (Mauritius) was captured by an expedition from India in 1810, and there exist projects between Napoleon and Admiral Decrès for the despatch of a conjoint expedition to India in 1805.

More interesting and more serious was the mission of General Gardane to Persia in 1807. Since 1804, by means of various emissaries, Napoleon had been in touch with that State, and in May, 1807, he cause Maret to sign the Treaty of Finkenstein with Mehemet Riza. By that treaty Napoleon recognised a Persian occupation of Georgia, and undertook to force the Russians to evacuate that country and other Persian provinces. He promised guns, muskets, and officers of all arms to reconstitute the Persian Army. Persia undertook to break with England and declare war with her ; to force the Afghans to join with her against England ; to receive a French squadron in Persian harbours, and to welcome any army which Napoleon might send overland.

This was followed up by the arrival of General Gardane, who reached Teheran in December, 1807, with a considerable staff, amidst great enthusiasm. The Treaty of Finkenstein was ratified. But the Treaties of Tilsit had been already signed by Napoleon and Alexander in July, and there was no mention of Persia in them. The Russians, rejecting Gardane's offer to mediate, took the field again and beat the Persians in November, 1808, when the prestige of France and of the French Mission was completely ruined. Persia had been sacrificed to Napoleon's other ambitions.

The Minute of April 12, 1807, from Napoleon to Talleyrand, and the Emperor's instructions of May 10 to General Gardane, make very instructive reading. Every precaution was taken for Gardane's rapid and secret journey, via Constantinople. The general and his staff were instructed to send in frequent and full reports, in duplicate by different hands, on a country 'sur lequel il n'existe aucun renseignement positif.' The Persians were to be made to regard the Russians as their natural enemies and to be aroused to make new efforts. Persia was to make a powerful diversion in the Caucasus. Persia and the Porte were to concert their operations between the Black Sea and the Caspian. Persia was to favour all French operations against England, and was to be told that if she did not she would become an English province. A French expedition, it was hinted, might be sent via Aleppo overland, or round the Cape to the Persian Gulf, and all reconnaissances and preparations were to be made for either event. The guns and muskets were to be paid for, because, as Napoleon said, if the Persians paid it would show that they were prepared to make use of the arms. Gardane was ordered to get in touch with our enemies in India. The object was described as an alliance between France, Turkey, and Persia, for the purpose of an advance on India, or, alternatively, as a scheme for the assembly of auxiliaries against Russia. The bait was Georgia, and the Persians swallowed it greedily.

The above details explain the reasons for the success and failure of some of our missions to Persia at this time. The mission of John Malcolm to Persia in 1800 preceded Napoleon's advances. It was mainly designed to effect a diversion, which would compel Zeman Shah to leave India. It was magnificently equipped, and Malcolm's presents were princely. The Shah of Persia bound himself not to permit the introduction of French influence and to harry the Afghans should they advance into India; but as these arrangements were, for reasons which are obscure, not formally ratified, nor their importance perhaps recognised by the British Government, there is some excuse for the Persians in their subsequent agreements with Gardane. They thought that we had deserted them.

After the peace of Tilsit Lord Minto, then Governor-General, sent missions to Lahore, Kabul, and Teheran to set up barriers against a Franco-Russian advance upon India. The mission of Elphinstone to Shah Shuja's camp at Peshawar in 1808 was not productive of political consequences. Malcolm was again selected for Teheran. He found French influence predominant, as the eyes of the Persians were not yet opened. There was another British envoy sent to Persia at this time—namely, Sir Harford Jones, who had been appointed by the home Government and the Court of Directors of the East Indian Company, and Malcolm returned to India in disgust. But the way of Sir Harford Jones was smoothed when the Persians discovered how Napoleon had treated them, and the French alliance was abandoned by Treaty, confirmed by Lord Minto. Any danger which there might have been from Russia naturally terminated when this Power and England became once more allies against Napoleon, and good relations between the two continued for many years after 1815.

These first relations between the British in India and the minor States beyond the Suleiman Mountains show in outline that India, Turkey, Russia, Afghanistan, Persia, and any enemy of England or Russia who is able by his alliances or conquests to gain a footing in this part of the world, all play certain *rôles* in frontier diplomacy and war. No one of them can be disregarded. If Paul had lived, the history of the nineteenth century might have been very different. We were not prepared for such a combination as that of France and Russia, and, as Malcolm's Treaty was not taken up, Napoleon was able to tie Persia to his interests before we were alive to what was happening. We were, in fact, forestalled, and the situation was only eased for us in the East because Napoleon sacrificed Persia for Alexander and lost them both. It was Tilsit rather than any forethought or efforts of ours which brought us through the years of 1800-10 unscathed.

1810-1857

In 1809 Shah Shuja, the Afghan ruler, was driven from his throne, and took refuge in British territory. In 1833-34 he made an expedition into Afghanistan to recover it, with British connivance, but was beaten by Dost Mahomed, who now ruled over North-Eastern Afghanistan, and must have regarded our approval of Shah Shuja's adventure as an unfriendly act. Three of the Dost's brothers held Kandahar, while Herat was in possession of his enemies, the Saduzai clan of the Duranis. The Sikhs had taken Peshawar, and had held it against Dost Mahomed's assault. That was broadly the position when Lord Auckland reached India in March, 1836.

The British friendship with Russia had cooled, and England was in the throes of an acute attack of Russophobia. Sir Robert Peel's Government had fallen in April, 1835. Just before its fall Lord Heytesbury had been appointed Governor-General, after having been Ambassador at St. Petersburg, where he had been on good terms with the Tsar Nicholas. This was enough for Palmerston, who was a distinguished victim of the prevailing disease and Foreign Secretary in Melbourne's Government. The new Ministry, in spite of the protests of the Court of Directors, cancelled Lord Heytesbury's appointment, and selected Lord Auckland in his place.

Waves of Russophobism from London soon began to break on the shores of India, and Auckland was entreated by the Government at home to raise a barrier against Russian encroachments. The nature of the barrier was not defined. He began by sending Alexander Burnes to Kabul, nominally on a commercial mission. Burnes was no novice in Central Asian travel, and soon established cordial personal relations with the Amir. He was probably right in believing that Dost Mahomed desired our friendship and good offices, but the Dost's desire to recover Peshawar, though supported by Burnes at Simla, naturally could not be accepted by the Governor-General. The Sikhs were still our faithful friends, and had won Peshawar in fair fight. We could not afford to advance into Afghanistan and leave an unfriendly Punjab behind us. Burnes had no real powers to treat and no proofs to offer of British friendship.

By May, 1838, Lord Auckland had recalled and thrown over his envoy, and with William Macnaghten at his elbow had decided to restore Shah Shuja to the throne of Kabul. A tripartite Treaty was made with Ranjit Singh and Shah Shuja to this effect. A public manifesto of a questionable character by Lord Auckland, denouncing Dost Mahomed, followed on October 1, and preparations for war began. Seldom, if ever, had we entered upon war with less provocation on the other side. The campaign ran its terrible and dramatic course. Shah Shuja was, indeed, carried by the British arms to Kandahar and Ghazni, and then to Kabul, which he entered in August, 1839. But here every political and military mistake that could be made was made. The garrison left behind was too weak, and affairs at Kabul were grossly mismanaged. Retreat under the most discouraging circumstances became inevitable. The Afghans broke their word to us, and out of a small force of 4,500 fighting men and 12,000 followers who attempted to join Sale at Jalalabad, Dr. Brydon alone reached that heroic garrison. Nott, Pollock, and Sale fought several battles to reopen the road to Kabul and re-establish our military reputation. We had begun the war to replace Dost Mahomed by Shah Shuja, and we ended it by replacing Shah Shuja by Dost Mahomed after a loss of 20,000 lives and 15 millions of money.

It was our first introduction to Afghanistan, and it should have conveyed salutary lessons. It was shown that an Afghan ruler supported by foreign bayonets became the most unpopular person in his country. It was shown that a British Mission or a weak British detachment at Kabul were not safe, owing to the fanatical hostility of the people. It was shown that when the British go into Afghanistan they must go there not as partners but as principals, prepared to carry things through with a high hand and with their own resources against any enemies. The difficulties of the country were to some extent exposed. Though the Afghans were not able to withstand our disciplined troops on any sort of fair fighting terms, they nevertheless displayed fortitude and resolution, and gave proofs of a spirit of independence beyond anything that we had experienced in India. We found it easy enough to enter Afghanistan, but hard to get out of it. The disastrous retreat from Kabul was the worst blow that our prestige had ever received in Asia. We did nothing in Afghanistan to raise those barriers against Russia which Melbourne's Government had vaguely required and Lord Auckland had as vaguely attempted to erect. By appearing in arms as the enemy of

Afghanistan we established a record not yet exhausted of its evil consequences, and we almost compelled an Afghan ruler, wnen threatened by us, to appeal to Russia for protection.

We found that our greatest danger beyond the mountains was not the shadowy figure of the Russian Colussus, but rather our own foolish fears and the confederacy in folly of a British Government and the Government of India. Nothing can excuse the course which Lord Auckland took when he dug out the wretched pensioner of Loudiana and tried to set him up in opposition to a man like Dost Mahomed. The Dost's later warning to us to let pretenders for the Afghan throne fight things out among themselves was a wise one, and if it was hard to follow owing to the extraordinary number of sons which Afghan rulers ordinarily left behind them—Dost Mahomed himself left sixteen —and also owing to the time which sometimes elapsed before one or other established himself securely upon the throne, it was always certain that our fussy interference in Afghan habits on the death of an Amir brought us more loss than profit.

Dost Mahomed himself had such an obsession for the recovery of Peshawar that he agreed, as the price of its session, to join in against us during the Second Sikh War. The Punjab was reeking with disaffection in 1848, but Sir George Lawrence might have kept the Sikh garrison of Peshawar faithful had not Dost Mahomed's brothers stampeded it. Afghans took part in the fighting of that campaign up to and including the battle of Gujerat, and when the Sikhs laid down their arms the Afghans were hunted back to their country by our cavalry and fled ignominiously through the passes. That score was against Dost Mahomed, and it seems to have been a lesson to him.

In 1856 we were at war with Persia, and a Persian army attacked and captured Herat. An expedition was sent to the Persian Gulf under Sir James Outram, and, after several successful combats by our troops, the Shah renounced all claims to Herat and withdrew his troops from Afghanistan. Dost Mahomed's action contributed to this success, and Sir John Lawrence's dealings with the Amir between 1854 and 1857 were largely the cause of the friendship of the Dost to us during the Mutiny. By the Treaty of 1855 contracted by Lawrence and Herbert Edwardes we engaged 'to respect those countries of Afghanistan now in his Highness's possession, and never to interfere therein,' while Dost Mahomed engaged never to interfere in our territory and 'to be friend of our friends and enemy of our enemies.' When the Persian War began Dost Mahomed had taken possession of Kandahar. He was then 70 years of age. The action of Persia had rendered a further Anglo-Afghan discussion necessary, and the Dost met Sir John Lawrence at Jamrud in January, 1857.

The question between Afghanistan and Persia is always not only one of this world but of the next, for Sunnis and Shiahs cannot easily unite. The Amir offered to send 50,000 men and 100 guns to Herat, or more if we wished it, and his sons asked for an important consignment of arms and ammunition and a large sum of money if such an expedition were undertaken. But for a defensive war the Afghans asked for 4,000 muskets and a lac of rupees a month while the war lasted. These last requests were granted, but our right to depute British officers to see that the money was properly applied at Kabul was strongly resisted. It was pointed out by the Afghan Sirdars that the people would not like it, that their national and religious feelings would be outraged, and that they would think that the days of Shah Shuja had returned. They asked that a native vakil should be placed at Kabul, and this was wisely agreed to.

The articles of agreement were signed after long palavering, and the Amir declared in the end that he had now made an alliance with the British Government, 'and come what may I will keep it till death.' He was as good as his word. Canning, Lawrence, and Herbert Edwardes thus secured us on the side of Afghanistan before the Mutiny began. Had bad policy made Dost Mahomed an enemy, our men on the Ridge before Delhi could never have been strengthened as they were by Lawrence from the Punjab. When a crisis in India and a hostile Afghanistan happen to coincide there is a prima facie case for the belief that bad policy reigns in our councils.

1857-1923

Elgin, Lawrence, Mayo, and Northbrook succeeded Canning as Governor-General in turn, and the external relations of India were wisely conducted until the close of Northbrook's administration. Dost Mahomed died in 1863, and it was not till 1868 that his son Shere Ali established himself upon that most thorny of seats, the Afghan throne. We had more or less made up our minds since 1857 to stop, as the Sikhs had stopped, at the base of the mountains on the N.W. Frontier, to leave the tribes their nominal independence, to respect the independence of Afghanistan, and to interfere with it as little as possible. It was, in the main, the Lawrence policy, so often travestied since by men who do not understand it, and it was strengthened in 1869 by Gortchakoff's pledge that Afghanistan should be considered to lie outside the sphere of Russian influence, and by an agreement with Russia in 1873 that the Oxus should be the northern boundary of Afghanistan.

Looking back, many of us nowadays are prepared to admit that the progress of Russia towards Afghanistan on one side, and our progress across Hindostan to the mountains on the other, were due to similar causes, and need not be attributed to deep-laid plans of aggression on either side. But few people in England had the coolness and the perspicacity to take this view at the time, and with every step forward taken by Russia our animosity increased and the anxiety of Shere Ali became more poignant. To the McNeills and David Urquharts of the thirties there succeeded the Rawlinsons, the Bartle Freres, and the Burnabys of the sixties and the seventies, and, while opinion at home became alarmed,

Northbrook could do little more than give Shere Ali vague promises of support. A Whig Government could not afford to repeat the Auckland fiasco.

The Amir's desire that we should consider Russia our enemy if she violated his frontier was comprehensible. So also was our view that there might be the greatest difficulty in deciding which party was the aggressor in a frontier brawl. It was a serious matter to plunge two great Empires into war over a frontier collision, in which the forces on both sides were apt by nature to follow a word by a blow. So we, in effect, did nothing, and it must be accounted unfortunate that statesmanship did not discover a formula to calm the Amir's legitimate fears after the Russian occupation of Khiva in 1873.

In 1874 Disraeli became Prime Minister and Salisbury Secretary for India. Russophobia resumed its old sway as in 1836, and while a stronger foreign policy had figured in the party programme at the General Election, circumstances combined to make Afghanistan the region where it was first applied. The remedy in favour at home was Rawlinson's plan of sending a British Resident to Herat first and then to Kabul. The first part of this programme was strongly pressed on Northbrook by Salisbury in January, 1875, and the whole remedy later in the year. Against this pressure Northbrook firmly stood out, supported by his Council, and quoted the opinion of frontier officers against it. He also warned Salisbury that the policy advocated was likely to bring on another unnecessary and costly war in Afghanistan. But the mind of the Government was made up. They were set upon a forward policy, and Northbrook, realising that he was not the man to carry out a policy with which he entirely disagreed, resigned in dignified silence. He was succeeded early in 1876 by Lytton, who was specially selected to carry out the strong policy of the Government, and was scarcely expected to have any policy of his own.

Things moved rapidly after Lytton's arrival in India. He

soon had his Council firmly in hand. But neither his letters to Kabul, nor a conference at Peshawar in January, 1877, could induce the Amir to waive his rooted objections to the presence of a British envoy at Kabul. Shere Ali's objections were the same as those of Dost Mahomed and his Sirdars in previous years, and while Lytton's communications became more minatory and unsympathetic, Disraeli stood squarely behind him and backed him up.

The Russo-Turkish war of 1877 affected the issues. We had taken up a strong pro-Turkish attitude in that war, and Russia, as a legitimate political counter-stroke, had directed some small columns to advance in Central Asia to vex us. Things came to a head in 1878 when the Russian General Stolietoff reached Kabul in July. Thereupon Lytton, with his Government's approval, demanded the prompt reception of a British envoy, and sent Sir Neville Chamberlain and Major Cavagnari to the frontier for this purpose. The mission reached the Khyber in September, and was refused admission to Afghanistan by the Afghan commander of Ali Musjid fort. An ultimatum followed, due for reply by November 20, and as no reply was received, the Anglo-Indian forces crossed the frontier in three columns and defeated the Afghan forces in their front.

Shere Ali fled to the north, and soon died, leaving his son Yakub Khan to make the best terms he could. A treaty was signed with him and Gandamak in May, 1879, and it gave us the right upon which we had insisted so pertinaciously to send a Resident to Kabul. The troops were withdrawn, and Major Cavagnari, with an escort and a few companions, reached Kabul on July 24. But on September 6, mutinous Afghan troops broke out, overcame the escort after a gallant resistance, and massacred the Mission. Our troops again advanced. Roberts occupied Kabul by the Kurram route after successful engagements, but met with difficulties later owing to risings of the people, which confronted him with numerically very superior forces. He was reinforced, and dispersed the gatherings, but we stood in a difficult position, for we had not the means to occupy and control the whole country, while to retire was to leave the situation worse than we had found it. But the Russian pensioner, Abdurrahman Khan, Shere Ali's nephew, fortunately appeared on the scene at this time, and to him Lytton wisely confided the throne, which was accepted. We then withdrew, and the victory of Roberts at Kandahar over Ayub Khan, who had marched from Herat and beaten a small Anglo-Indian force at Maiwand, closed the campaign.

Ripon succeeded Lytton, and in July, 1880, Abdurrahman was formally recognised as Amir. We promised him protection against aggression by any other Power if he would follow our advice in his foreign relations, and, while we granted him a subsidy of twelve lahhs a year, we did not insist upon a right to send a British envoy to Kabul. Abdurrahman ruled Afghanistan with an iron hand, and obtained complete mastery over it. During Dufferin's Viceroyalty the Russians occupied Merv in 1884, and in the following year a collision between the Russians and the Afghans occurred at Penjdeh while the British and Russian representatives were engaged upon the delimitation of the frontier line. War seemed near, but the statesmanship of Dufferin and Abdurrahman, who happened to be together at Rawul Pindi at the time, fortunately prevented it. The Boundary Commission completed its labours, and an agreement made at St. Peterburg in 1887 closed a dangerous episode.

THE RISING OF 1897

During the administrations of Lansdowne and Elgin the forward school resumed a certain amount of control in India. They desired to occupy the tribal territory, to delimit our political frontier with Afghanistan, and to extend our strategic railways. These ideas were not acceptable to Abdurrahman, who wished the tribes to remain independent, but in 1893 Sir Mortimer Durrand secured the consent of the Amir to the delimitation of the Afghan frontier, and this was rounded off two years later by a delimitation in the Pamirs.

But the acts of the forward school from 1890 to 1897 had to be paid for. During those years we had occupied the Samana and the Kurram valley, had opened the Gomal, placed posts in the Sherani country, penetrated the Tochi, established garrisons at Wana and Miranshah, had been active in Dir and Swat, and had quartered troops at Malakand and Chitral. These acts were in the main justified from a military point of view, but they aroused fierce resentment among the tribes, and were very disturbing to the Amir. The part played by the latter in the subsequent events has never been fully recorded, but without Afghan connivance the rising which occurred so suddenly and simultaneously from the Tochi to Chitral could hardly have happened. This rising proved a great strain upon us, and led to expeditions by larger forces than we had ever employed in tribal territory. We entered Tirah, so long judged inaccessible to us by the Afridis, but we did not attempt to occupy it, and affairs ultimately settled down again.

The Curzon Policy

To the forward policy Curzon paid little tribute during his term of office in the East. He kept friends with the Amir, and opposed all interference with the tribes and new commitments in their territories. By separating the North-West Frontier Province from the Punjab in 1901, and by making the Chief Commissioner the Agent for the Governor-General, he secured a firmer grasp on frontier policy by the Viceroy. He was active in Persia, and settled long-outstanding difficulties by the Convention respecting Persia with Russia of September, 1907. Let us give to the old Imperial Russia this credit, that whenever we approached her and dealt fairly with her on Central Asian questions she dealt fairly with us. It was less Russia than unworthy fears of Russia that were at the bottom of nearly all our mistakes and disappointments.

The Latest Phase

The murder of Habibulla, the succession of Amanulla, his attack upon us in 1919, and the Treaty of 1921, are all too near to us for anyone to forget them. The new Amir remains an uncertain quantity, and his refusal any longer to follow our advice in his foreign relations absolves us from responsibility for the defence of his country. There are rumours that he has made treaties with Russia, Turkey, Persia, Italy, and perhaps with France. If so, these are new facts which have to be taken into account by him and by us, and they are complicated by his inability or dislike to hand over murderers who have taken refuge in his territory, and by our present unnecessary and hampering entanglement in Waziriland.

The defence of the North-West Frontier of India is a political and military problem of the utmost complexity rendered all the more difficult owing to the uncertainty of such factors as religious fanaticism and external intrigue, the variable policy of British Governments, and the idiosyncrasies of individuals, whether British or Afghan. We have never gathered up all the lessons of the past century, nor have constructed from them a code of frontier doctrine consecrated by experience.

But from a broad point of view the Lawrence policy holds the field, and its general wisdom has been proved as much by our success when we have followed it as by our failure when we have departed from it. Certainly, we have to advance with the times and not to remain in old ruts when progress shows that they no longer serve us. But the Second Afghan War, the rising of 1897, and contemporary events, have all shown us that Lawrence had drawn the right deductions from 1836-42, and those Viceroys who have departed from them are reckoned among our greatest failures.

Lawrence's principles were that we should halt at the base of the mountains, preserve friendship with Afghanistan, respect the independence of the tribes, avoid commitments in tribal territory, and rest content to create strong forces behind the mountain barrier, ready to intervene promptly and effectually wherever our vital interests might dictate. The whole of the lessons of a century behind us point to the wisdom and statesmanship of those principles which have gained new strength from the events which have happened since Lawrence's death, and are reinforced to-day by the character of our military establishments and the nature of the great experiment in self-government to which we stand committed in India. It is surely time that these principles should be understood and followed by Governments at home and in India, as well as by the General Staffs and Staff Colleges of the British Empire.

CHAPTER XVIII

Afghanistan

THE AMIR'S RECORD

H.H. the Amir Amanulla Khan—His Two Predecessors—Abdurrahman One of the Strongest and Wisest Rulers that Central Asia has ever Produced—Habibulla Follows in His Father's Footsteps—Murder of Habibulla—Amanulla Succeeds—He Declares Afghanistan to be an Independent Kingdom—Our Control of the Foreign Relations of Afghanistan Ceases—The New Amir Makes War Upon Us—Preaches a Jehad—Marches on India—His Army is Attacked and Beaten—Headlong Flight—The Amir Asks that Hostilities May Cease—The Treaty of 1921—Amanulla Turns to Reforms in Afghanistan—His Dubious Attitude at Jalalabad in February, 1923—The Murders of Majors Orr and Anderson by Shinwari Afghan Subjects—Foreign Legations Established at Kabul—Our Desire that Afghanistan Should Be Strong and Independent—The Condition is that Her Ruler Should Be a Friend.

HIS Majesty the Amir Amanulla Khan has reigned over his kingdom for more than four years without leaving upon us a very clear or a very satisfactory impression of his individuality or his policy. This is largely due, no doubt, to the rarity of the occasions upon which competent observers visit his country, to the general inaccessibility of Afghanistan, and to the ignorance of most Europeans of Persian and Pushtoo. Our Legation at Kabul is, of course, in a privileged position, and officially we are doubtless well informed, but the public would like to know more of a ruler who started his career by lightheartedly making war upon us in India, and by severing the links which, since the fall of the Amir Shere Ali, had bound the Afghanistan of Abdurrahman and of Habibulla to the Indian Empire.

Amanulla's two immediate predecessors were very good friends of ours, although we often had delicate and difficult negotiations with them, and Abdurrahman, in particular, was one of the strongest and wisest rulers that Central Asia has seen in modern times. He made modern Afghanistan, subdued all his rebels, largely replaced the old tribal levies by regulars, ruled like the autocrat he was, and bent the Afghans to his will. His son Habibulla followed in his father's footsteps, and relied, as his father had done, on the English alliance. Thanks to his prudence, peace was preserved in Afghanistan throughout the World War, even though, in 1917, and 1918, the Bolshevists were on the frontier and ripe for mischief.

But when, in February, 1919, Habibulla was murdered in his bed, and his son, Amanulla, succeeded at the age of 29, after his uncle, the Sirdar Nasrulla Khan, had made an unsuccessful attempt to gain the throne, the scene suddenly changed. Amanulla, without a moment's delay, declared Afghanistan to be a free and independent kingdom, and under the heavy weight of Imamat and Amarat placed the crown upon his own head. He claimed, that is to say, both religious and political supremacy, and informed his 'nation, with a nice sense of honour,' that all rights of government claimed by other independent Powers throughout the world should in future be possessed in their entirety by Afghanistan.

That, of course, changed the situation profoundly, though at first the Viceroy, Lord Chelmsford, wisely feigned to ignore the implications of the announcement. Our control of the foreign relations of Afghanistan was suddenly at an end, and so was the obligation which rested upon us to defend Afghanistan if it were attacked, and to pay subsidies to the Amir. It was something of a shock, but on mature consideration the Amir appeared to be the loser over the transaction, and we were relieved from the disagreeable necessity of marching armies into Afghanistan to defend its independence.

But we very soon found that there was much more behind, for from a Firman sent out under seal of the Amir in various directions we soon learnt that he had been filled up by Indian seditionists, through the intermediary of the Afghan postmaster at Peshawar, with ridiculous notions about the situation in India, and that he was prepared to come down from his hills as champion of a people who, he imagined, were distressed and persecuted. In fact, in special Durbar, he preached a Holy War, and sent inflammatory messages throughout his kingdom, into India, and amongst the independent tribes. He marched his regulars towards India, attempted to raise hordes of Ghazis, and planned attacks from Dakka, Khost, and Quetta upon the Indian frontier. India and the frontier tribes were to rise at the same moment, and, altogether, we were back in the good old days of the jehad, with Afghan lashkars in every pass, and with blood and rapine for the watchword. Whatever we may have thought of the new Amir, we could not deny that he was thorough.

The results did not respond to the intentions. The hordes of Ghazis marched towards but did not debouch upon, the frontier. The independent tribes, restrained by the firm hand of the late Sir George Roos Keppel, did not rise. India remained quiescent, and only stirred to the extent of sending formidable forces to the border. The Khyber Rifles, it is true, became unsteady, and had, in the main, to be discharged, but Mohmands, Waziris, Zaimukhts, Orakzais, and Afridis kept very cool, and in general the Pathans sat upon the fence and waited to see what would happen. From Dera Ismail Khan, Hazara, Kohat, and Bannu the reports were that all was quiet. The Afridis considered that the Afghans had betrayed them in 1908 and in 1897, and waited for the armies of 'the God-granted kingdom' to lead off.

They did their best. But on May 9 they were driven from their positions in our territory near Landi Khana, Dakka being bombed with good effect; and on May 17 they were attacked on the ridge west of Dakka and thoroughly well beaten, while Basawal and Jelalabad were well bombed.

From the latter town, which was the Afghan advanced base, there was a headlong flight, whereupon the watching tribesmen most unkindly looted the arms, stores, and food from their would-be protectors. The news flew round and had excellent effect, while Kabul was visited by an aeroplane, and the wails that its visit caused showed that it had struck home. In a wonderfully short space of time the whole Afghan attack collapsed, and the Amir asked for a cessation of hostilities. It was granted to him, as he was bluntly told, on account of his youth and inexperience, and because he had been led astray by bad advisers; but he had to listen to a lecture from Lord Chelmsford which did not mince matters before he was permitted to escape from further consequences of his folly.

This war had, of course, abolished all our treaties with Afghanistan, and though there was no desire to revert to the old arrangements, it was necessary to establish neighbourly relations, to define the frontier once more, and to agree upon Customs, trade agents, posts, Legations, and Consulates. This was done by the Treaty signed at Kabul on November 22, 1921. There is one interesting political point, not in this Treaty itself, but in a letter from the British representative at the time to the Afghan Foreign Minister, to the effect that the British Government 'entertains feelings of good will towards all the frontier tribes, and has every intention of treating them generously, provided they abstain from outrages against the inhabitants of India.'

This was an important point. On racial and religious grounds the new Amir considered that the independent tribes on our side of the Durand line had claims to his protection, and he endeavoured to lay stress on the fact in the Treaty. That we could not admit, and never have admitted, for once we admit claims of a foreign Power to protect people within our political frontiers we by so much abandon sovereignty over them. We cannot describe

Amanulla's claims as altogether new. Both Abdurrahman and Habibulla—the latter especially—were thought at times to have encouraged the tribes, but on the other hand was the fact that we did not always keep our frontier banditti in order, and the Amirs of Kabul were in some sense bound to humour them, or they might have proved as troublesome to Kabul as to us. In part, perhaps, we can ascribe our new policy in Waziristan to the reminder which Amanulla gave us while negotiating the Treaty of 1921.

The events of 1919 had compelled Amanulla to recognise our power and the futility of the information upon which his aggressive and wholly unprovoked attitude had been based. He had the wisdom to abandon a policy which threatened to destroy his prestige in his own country, even if it did not lead to consequences still more ruinous to him. He turned, and wisely turned, to the task of the administration of his own country, where much remained to be done, and he gradually won the approval of the Anglo-Indian Press, and seemed disposed to realise where his true interests lay. Though invited to attend, and it is said to preside over, a meeting of Islamic Powers initiated by Angora, he is understood to have refused, and he looked like broadening down into a responsible and efficient head of a neighbouring Government.

This continued till February 1923, when, during one of his usual winter visits to Jelalabad, he appeared to suffer a relapse in good intentions and to pose once more as the protector of the frontier tribes. He dressed himself like a Pathan Highlander, spoke Pushtoo instead of Persian, and, collecting representatives of the tribes about him, uttered words construed by the border warriors in a sense favourable to their interests. Some thought him hostile to us, and some did not. There is an element of equivocation in his addresses, and a lack of clarity, so it is not to be wondered at that some frontier outrages which followed hard upon his addresses were put down as a natural consequence of his incitements. This was especially true in the case of the murders of Majors Orr and Anderson in the Khyber, since the murderers were Shinwaris, and Afghan subjects. The arrest of these men by the Amir was a satisfaction to us. Their prompt trial, conviction, and punishment, if guilty, is still expected. We shall see, and the Amir will once more observe that people who play with fire on the border often burn themselves.

Our Legation at Kabul is in the competent hands of Colonel Humphrys, who speaks both Persian and Pushtoo, and will no doubt be able to give the Amir a great deal of information which in the past he has lacked. The Amir has a representative here, namely, Sirdar Abdul Hadi Khan, who has every opportunity of doing good owing to the close personal intercourse which his relations with the Foreign Office place at his disposal, and various Powers—France, Italy, and Russia—now have Legations at Kabul.

So far as our public here are concerned, the new Amir is under observation, and definite conclusions concerning his character and his policy are not yet formed. The treaty with him is almost non-political in character. His attitude in 1919 deserved the strongest reprobation, and there are some who regret that the punishment was not more exemplary. But Lord Chelmsford and his Council took a statesmanlike course when they were content to ascribe Amanulla's outbreak to youth and ignorance, and it would not take a great effort on the Amir's part for us both to resume confidential relationship, provided he were ready to make that effort.

Our object is to ensure that Afghanistan is strong and independent, and as that must also be the policy of every Amir who is neither mad nor swollen-headed, all the elements for good understanding exist. The fate of Bokhara and other territories now under the Soviet sway, as well as recently-reported Bolshevist military movements in Central Asia, should inspire at Kabul a wholesome distrust of a régime which would sweep away Amanulla and all his dynasty into the dust of oblivion were it to gain a foothold in Afghanistan. We ask nothing from Afghanistan but neighbourly relations, and have not the slightest desire to interfere at all in that country.

But a postulate certainty is that the Amir should neither act nor intrigue against us, whether with other Powers on within the borders of India. Were he to do so, the consequences would cause him great astonishment. In our past little expeditions into Afghanistan the bases of our power were far away from the frontier. Now they abut upon it, and things which were difficult in the old days are easy now, while the equipment and armament of our modern armies make resistance hopeless both on land and in the air.

His Majesty Amanulla Khan, Amir of Afghanistan and its dependencies, is an autocrat, and one of the very few left in the world. Afghanistan has not an enemy in England or in India. We wish her well, and can aid her to civilise her people and to make them prosperous and wealthy. But a condition is that we have confidence in the friendship of her ruler, and that we hear no more of those hostile acts and dubious designs which have marred the early years of Amanulla's reign, and have gone near to ruin irretrievably the dynasty of the Barakzais and to imperil the high destinies of Afghanistan.

CHAPTER XIX

Frontier Tragedies

The Frontier Tragedies of 1923—Great Characters of the N.W. Frontier of India—Men Not Systems—A Highland Precedent—Story of the Kohat Outrage in 1923—The Villains of the Piece—Heroines and Heroes—A Story of Real Life on the Frontier.

MOST home-staying people can have little idea of the dangers incurred by the heroic Mrs. Starr in her recent successful mission of rescue across the frontier. This gallant lady went voluntarily and willingly into a district where few white people travel who are not tired of life, and among tribes who, while possessing great martial qualities which almost endear them to soldiers, are certainly among the most treacherous and bloodthirsty races of the world. It was a triumph of character, and must have sent a thrill of admiration throughout India and amongst all old Anglo-Indians at home.

We can afford to spare a word of commendation for the devotion of the ressaldar who accompanied her, and for the courage of Sir John Maffey, who must have authorised the attempt, for had it failed the Chief Commissioner would have been bitterly criticised. But our chief admiration is necessarily for Mrs. Starr, who, so far as I can recall, is the first woman who has added her name to the *rôle* of the immortals on this frontier.

Probably most of us have studied the various systems of frontier management at one time or another, and many will share the opinion once expressed by Lord Curzon that 'there is no frontier policy capable of being framed that can be described as absolutely safe.' That is true, but the real glory of this frontier is the number of great men that it has produced, men who have not only given the best of their lives to their work, but have raised the renown of the British people to lofty heights by the greatness of their character, and by the courage, firmness, and justice which they have displayed in their dealings with the tribal territory and its almost unmanageable folk.

We throw our minds back and think of John Nicholson, Herbert Edwardes, James Abbott, Frederick Mackeson, John Jacob, Robert Sandeman, Lumsden, Cavagnari, Warburton, Roos-Keppel, and many others whose lives and experiences are the best records that we possess of this troubled land. In every case it is their character that attracts us, characters that were worth more than armies to us and saved us from many wars. There was a Commissioner at Peshawar—Major James—more than half a century ago, who expressed in a striking paragraph the influence of personality among Eastern races.

'It is to the straightforward, upright, and disinterested action of English gentlemen, and to the influence which higher mental culture never fails to exercise, rather than to superiority of fighting power and appliances that I attribute British supremacy in India, as well as the exceptional success of British rule in all quarters of the globe.'

That is the special quality of character for which our soldier-administrators and civilians on the North-West Frontier have ever been famous, and no amount of force can quite make up for its absence. When the example comes from such choice spirits it is taken up by the subordinates below, who become students in a great school; and so, when we want to discover the spirit which exists on the marches of India, we study these men rather than a policy dictated by London or by Simla.

What Abbott was to the Hazara tribes, Lumsden to the Eusofzais, Nicholson to the men of Pindi, Sandeman to the Baluchis, and Warburton to the Khyber folk, our present and less-known men are to the people whose affairs they manage. The great 'Nikkul-Seyn' was worshipped almost as a god, and it was with the utmost difficulty that the brotherhood of fakirs which worshipped him was suppressed. It was by their character first and then by their intimate acquaintance with the tribal men and politics, and by constant personal touch, that all these men won the confidence of their warlike and restless people. Nothing can ever be substituted for the influence of this personal character, and men may be ever so clever but will fail unless they possess it and have the necessary acquaintance with Persian and Pushtoo.

There was a debate on the N.W. Frontier in the House of Lords last year, when Lord Chelmsford said some very necessary things about the bombing of tribal villages without notice. This new system is as unlike that which I have briefly described as well can be. It is a policy of exasperation, and there has been nothing like it since the detestable policy after Culloden, which is still handed down to infamy in the old Jacobite songs. The independent tribes bear no malice for a good fight in which they are beaten, because they are warriors themselves and respect courage. So did Prince Charlie's men respect Colonel Gardiner after Prestonpans, because he had died a soldier's death on that day so disastrous for English arms. But this new expedient of bombing villages containing women and children, without notice, and of destroying the tribesmen's cattle is barbarous, stupid, and indefensible.

I wonder whether the Government of India have ever heard of the proposal for the settlement of the Highlands which Duncan Forbes submitted in 1738 through Lord Milton and the Earl of Islay to Sir Robert Walpole. It was to form regiments of Highlanders from the disaffected clans and to employ them on Government service. The proposal was described by Walpole as the most sensible plan he had ever seen. Had it been adopted the Rebellion of 1745 could hardly have taken place, but it was thrown out by the Council on the grounds that the Opposition would say that

Walpole designed to subvert the British Constitution and enslave the people of England !

How many Pathans from the tribes beyond our administrative border are now serving in the Indian Army can only be stated by an answer in Parliament, but there is so much and so curious similiarity between 1738 in the Highlands of Scotland and 1923 in the tribal territory that the suggestion seems worth examination. When Macdonald of Keppoch was asked in 1740 what the rent of his estate was, he applied that he could raise 500 men. It is imaginable that a tribal jirgah to-day would say much the same thing.

The recent outrage at Kohat, in its causes and in its results so far, affords us such an illuminating glimpse of real life on the North-West Frontier of India that it seems worth while to piece together the disconnected parts of it into a single story.

The murder of Majors Anderson and Orr in the Khyber is attributed to the Shinwaris, who have yet to be called to account for it ; but there is no indication that this affair had any connection with the Kohat outrage, which was the act of the Tirah Jowakis and the Bosti Khel sections of the Afridis, between whom and the Shinwaris there is not much love lost.

The villain of the piece at Kohat was a Bosti Khel Afridi named Ajab, whose ordinary home is only five miles distant from the Kohat cantonment, in a valley below and to the north of the frontier constabulary post at the kotal, on the Kohat Pass road. Ajab was supposed to be loyal, but after a large quantity of rifles had been stolen from the Kohat Police lines suspicion lighted on him, and his village was raided in a most successful manner one night by the constabulary. Not only were nearly all the rifles recovered, but certain articles were found in a cache implicating Ajab in the murder of Colonel Foulkes and his wife some four years ago.

Ajab himself was not caught, but the discovery meant outlawry and ruin to him, and he took refuge with the Tirah Jowakis, where there lived one Sultan Mir, who was the leader of the raiders who killed Colonel Foulkes, and resided at Khanki Bazaar, in the Tirah country, some twenty-seven miles from Kohat. The loss by Ajab of his home and his property, and the sarcasms of his women-folk, appear to have implanted in his mind the desire for revenge, and he is said to have gone to his shrewish mother with the Koran in his hands, and to have sworn to do something which would make everybody's flesh creep.

He had no difficulty in getting a handful of rascals to help him, and he was, of course, perfectly acquainted with the tracks between Khanki Bazaar and Kohat, and with the cantonment itself. So on a stormy night, which aided his infernal purpose, he broke into the bungalow of Colonel Ellis, who would probably have shared the fate of Colonel Foulkes had he not been absent on duty. What happened in the bungalow we do not know, except that one Shahzada, a fellow-ruffian of Ajab, killed Mrs Ellis, while Ajab himself was chiefly concerned with the abduction of her daughter.

What was his object? It happened that we were about to make a settlement with the Tirah Jowakis at Peshawar on April 17, four days after Ajab's raid, and that, as the settlement would have covered the Foulkes murder Ajab expected to be excluded from grace, and therefore sought to get something in hand in order to bargain for a pardon. That would be quite in accord with a Pathan's manner of thinking, and there is little doubt that this was in his mind.

Ajab and his charming friends, half carrying and half driving Miss Ellis, crossed the polo ground, turned east towards Khushalgarh and reached the hills east of Kohat Kotal and south of the Peshawar-Kohat road, where Ajab was on his native heath. There, in some hiding-place, the party remained during the day of April 14, and at nightfall resumed the march, moving always by night and lying close by day, until on the sixth day they reached Khanki Bazaar and accounted themselves safe. So cunning were they that not a soul is supposed to have met them, and the Afridi and Orakzai jirghas could at first give us no information whatsoever. This does not mean much on the border, however, because the law of hospitality is sacred if a fugitive claims sanctuary, and this is one of our difficulties in tribal territory.

Our local officers, meanwhile, had not been idle. Colonel Bruce, at Kohat, co-operating with Major Finnis, the Political Agent in the Khyber, since murdered by Wazirs on November 2, stirred up the tribes before dawn on the day of the murder, while Sir John Maffey promptly took charge of the whole arrangements. The guilt of Ajab must have been suspected, for by April 20 Khan Bahadur Kuli Khan, Assistant Political Agent in the Kurram, reached Khanki Bazaar, accompanied by an Afridi jirgah. The action of this Agent was highly praiseworthy. He was on a cold scent, and the gang strenuously denied all knowledge of the abduction, yet before Mrs. Starr, and her escort Ressaldar Moghal Baz Khan, reached Khanki Bazaar, an admission had been extracted that Miss Ellis was in one of the usual fortified Afridi villages a few miles away.

Mrs. Starr, unhappily for herself, is not unversed in frontier treachery. Five years ago her husband, Dr. Vernon Starr, was Superintendent of the Church Mission Hospital at Peshawar. On a Sunday morning before dawn, in March, 1918, three men carrying lanterns knocked him up on the pretence that a case of sickness required his immediate help, and as the good doctor opened the door he was stabbed to death. He had done splendid work, and so has Mrs. Starr done since ; it was not in ignorance of her danger that she promptly offered to go to the help of her fellow-countrywoman.

Mrs. Starr and her escort, on reaching the Orakzai country, received a clear intimation from the Mullah Mahmud Akhunzada, who lives at Khanki Bazaar, that her presence was unwelcome. That advice would have turned back many people careful of their lives, but neither Mrs. Starr nor the ressaldar took any notice of it, and calmly continued their journey to the Mullah's fastness, and the representations of the three emissaries of the Government finally induced the Mullah and Ajab to transfer Miss Ellis to the Mullah's house, where the ministrations of Mrs. Starr must have been as welcome as unexpected.

It seems to have been touch and go on April 22, for Ajab and Shahzada, while with the Mullah, learnt that an Afridi rescue party were attacking their village, and became infuriated. Shahzada laid hands on Mrs. Starr and forced her out of the house, while threats were made to murder both ladies. At this critical moment the Mullah rose to the occasion. It was something of a revelation. When we are on this frontier we are not in the habit of thinking much of Afridi Mullahs. In fact, the Afridis have a repute of not thinking much of Mullahs themselves. The frontier story goes that the Afridis were once teased by some neighbouring clans that there was not such a thing as a ziyarat, or sacred shrine, in the whole of Afridiland. The Afridis took the chaff to heart, and after solemnly talking the matter over remembered that they had a Mullah somewhere, and so sent for him, murdered him, and set up a ziyarat over his bones. That is, at least, one way of founding a Westminster Abbey.

We must revise our opinions about Afridi Mullahs. The Mullah Mahmud Akhunzada, as I have said, rose to the occasion. Much incensed at the insult to the sanctity of his home and to the sacred laws of hospitality, he cursed Ajab, Shahzada, and company publicly, loudly, and lengthily. Appalled by these priestly damnations, the villains gave way, and on the morning of April 23 the two Englishwomen and their two most gallant Indian escorters made tracks with the greatest rapidity to Shinawara lest the effect of the Mullah's imprecations might wear thin.

The pluck, coolness, and endurance of Miss Ellis during this tremendous strain on nerves and body were indeed remarkable, while the self-sacrifice and devotion of Mrs. Starr have aroused general admiration. But Khan Bahadur Kuli Khan and

Ressaldar Moghal Baz Khan well deserve the honours bestowed upon them, and if we can find some way of acknowledging the efficacy of the Mullah's ecclesiastical anathemas it would perhaps create a precedent, but still be consoling.

In Sir John Maffey we evidently possess a frontier leader of the old stamp who has the moral courage to take a great responsibility on himself and to act with the rapidity of thought. If there is one thing that appeals to us on the administrative side in this affair it is the excellence of the Chief Commissioner's Intelligence Service, to which we must take off our hats, observing that the two parties sent to the rescue made a bee-line for Khanki Bazaar before the Afridi and Orakzai jirghas had any knowledge of the existence or whereabouts of the ruffians. The Sirkar, thank goodness, comes out of this business right well.

CHAPTER XX

Indian Frontier Policy

A New Commitment

A New Commitment: The North-West Frontier Province—Political and Military Schools of Thought—No Policy Thought Out—All Our Old Policy Aimed at Defence Against Russia—The Fall of Imperial Russia in 1917—The Independence of Afghanistan—An Army Sent by Us into Waziristan in 1920—We Accept Afghan Independence with Calm—Our Liability to Defend Afghanistan Ceases—The Waziristan Entanglement—A Partial and Permanent Occupation—Our Methods Combine all Mistakes that Can be Committed on this Frontier—Arguments Against the Course Taken—A Perpetual Source of Future Trouble. The Protected Border: The Khyber and the Bolan—Baluchistan—A Policy Needed for the Rest of the Border—Trans-border Commitments—Our Present Policy Will Provoke the Amir and the Tribes—Hadrian's Wall—The Frontier Constabulary—Present System of Border Protection—Factors of Success—What We Need to Make the Border Secure—Suggested Restoration of a Frontier Force—A Searchlight Needed—The Present Policy Conflicts with Our Assurances to the Tribes—The Precedent of 1897—A Clear Policy Needed Towards Afghanistan and the Tribes—Success Depends on the Character of the Men Whom We Employ on the Frontier. The Forward School: Policy and General Staffs—Waziristan—Road Construction—The General Question—Effects on Afghanistan.

Most people know that between our administrative border on the N.W. Frontier of India and our political frontier with Afghanistan there is a fringe of virtually independent tribal territory, filled with a warlike and well-armed people very jealous of their democratic liberties. With these people we have had many a fight, but we remain on pretty good terms on the whole, for martial races respect one another.

Though the whole frontier from the Bay of Bengal to the

north-west of Kashmir measures 1,400 miles, the part which concerns us most is the North-West Frontier Province, created in 1901, covering an area of 16,466 square miles, and containing 2,500,000 people. The Chief Commissioner of this important province acts in a dual capacity, for he is directly under the Government of India as Commissioner, and also control the political relations with the tribes as Agent of the Governor-General. The estimated population of the agencies and tribal areas is 2,770,666 souls. The Chief Commissioner administers the settled districts of his province through five Deputy Commissioners, and controls the agencies and the tribes through political officers at the Malakand—which includes Dir, Swat, and Chitral—and at the Khyber, the Kurram, the Tochi, and at Wana. At this moment the two last political charges are in the hands of the Resident in Waziristan.

It has been a moot question, ever since we took over from the Sikhs in 1849, how best to deal with the fringe of virile and interesting desperados along our border. We have had, both in statecraft and in strategy, our 'forward school,' our 'close-border' school, and our 'back to the Indus' school, to which we must now add a fourth, which has been invented on the spur of the moment to explain our present policy in Waziristan, and is at once 'half-forward' and 'half-back.' No school has converted the others, and none has been quite successful. We can eliminate the 'back to Indus' school, because it would leave our fellow-citizens of the North-West Frontier Province to the tender mercies of the tribes; but the other schools become predominant in turn, and no settled or regular policy has ever existed for long.

Lord Morley, when Secretary for India, once told me that he never looked further than a fortnight ahead of him, a remark open to a variety of constructions and comments. So far as I am aware, our civil and military pundits in India and at home have never sat round a table together and thought a policy out.

We find this very strange, because the only Empire in the past at all resembling ours was that of Rome, and the collapse of Roman rule in the middle of the third century grew out of the unhappy conduct of frontier defence at several places simultaneously. From the time when Russia crossed the Caspian and began gradually to push her railheads towards the Afghan frontier, until the fall of Imperial Russia in 1917, all our strategy on the frontier of India was directed to defence against Russia.

Soldiers gradually became convinced, from a close and long examination of the problem, that the correct strategy in Asia was to fight Russia on the Hindu Kush and the Helmund in co-operation with the Afghans, and all our preparations were directed to that object, both in India and at home. The tribal question became burning when we had agreed to help Afghanistan if her ruler would follow our advice in his foreign relations, and when Sir Mortimer Durand and the Amir Abdur Rahman had traced out the political frontier between India and Afghanistan in 1893.

It seemed to soldiers then, after this political frontier had been settled, that our proper course was to occupy the tribal territory up to the Durand line, and to bring the tribes on our side of this line under our direct administration and control, so that they should not be able to play tricks on our communications while we had armies in Afghanistan. I myself adhere to this view. The thing might have been done, but other ideas prevailed in political spheres, and, with the steady increase of tribal population and their acquisition of modern arms, it became a bigger job every year. The fact that, during the thirty years which have elapsed since the Durand line was laid down, no Viceroy has seen his way to occupy and administer the tribal territory says much for the difficulty of the undertaking.

Much could be, and has been, written upon this subject, but it is all ancient history now, except in so far that the result up to 1922 was that our preserve of bandits was maintained, and that the old system of raid and counter-raid continued. Expeditions, large and small, entered the tribal territory, had some real good fights which both sides enjoyed, and came out again, but, as a general rule, this territory and the virtual independence of its people remained unaffected. So, too, did the northern frontier of Italy up to the days of Augustus, and for much the same reason. Rome's frontier problem was as big as ours, and no one before Augustus ventured to tackle it.

In 1917 and 1919 two very important changes occurred. In the former year Imperial Russia collapsed, and in the latter our good friend the Amir Habibulla was murdered in his bed. His son Amanulla, immediately after his accession, declared Afghanistan to be a free and independent kingdom. It was his right, for our arrangements with each Amir were personal and not dynastic. He went to war with us, and was let off lightly owing to his youth and inexperience. Our control over his external relations ceased, and also our liability to defend his country from attack.

A third event happened in 1920, namely, a decision to send a strong Anglo-Indian force, eventually 45,000 strong, into Waziristan—for the ninth time, so far as I can recall—to chastise its people, especially the Mahsuds, who had thoroughly deserved punishment. The total Wazir and Mahsud fighting strength was estimated at about 50,000 men. An arduous campaign followed.

We accepted the declaration of independence by Afghanistan with calm. Certainly, there was an inconvenient side to it. Various nations—French, Italian, and Russian—sent Legations to Kabul, which then became a pawn on the world's chessboard. What the Latin foreign Legations find to do at Kabul no one has yet discovered, but a Soviet Russian Minister there has been caught trying feebly to bribe and arm the frontier tribes against us. His failure was due rather to want of means than want of ill-will. On the other hand, the ending of our liability to send armies into Afghanistan to help the Afghans was a decided relief, and considerably freed our hands. There was no longer urgent need on military grounds, to occupy the tribal territory between our administrative border and Afghanistan.

It certainly seemed that Soviet Russia was considerably weaker than the old Russia from a military point of view. The conduct of the Soviet troops in the campaign against Poland exposed their weakness, and any great advance against India—not against Afghanistan—from the Kushk-Termez lines seemed beyond their powers. Such advance must always be mainly a problem of railways and highly efficient organisation, in neither of which arts had the Soviet Government displayed any capacity at all. We expected, of course, that Trotsky's crew would bluff, rob, murder, and intrigue in what we once called the Khanates, but we supposed that Amanulla would see where his interest lay, and not call down upon himself the fate of Bokhara. However, these things might befall, we felt that we should have a much longer warning than formerly before any serious attack could develop against the borderland of India. The Hindustane Fanatic colony at Chamarkand has endeavoured, it is true, to spread communistic propaganda amongst the tribes of Dir and the Hazara border, but has not had much luck locally.

But then the unexpected again happened. We gradually discovered that the Government of India had not only sent an army into Waziristan, but meant to occupy it. A complete occupation was apparently found to be too great and costly a business, but our public have not been informed how matters passed at this particular stage. The nature of the eventual policy was not fully disclosed to the public at home until the text of Mr. Denys Bray's speech in the Legislative Assembly of India on March 5 of this year reached England. Then we saw that the Government of India had shied at the cost of complete occupation and had invented a new school of frontier political strategy, namely, the 'half-forward' school, and was endeavouring to prove to us what a wonderful invention it was. The speech at the time was locally described as stirring. It was. It reminded us of the witches stirring their cauldron of toil and trouble. It appeared that a large garrison was to be kept in Waziristan until a circular road, metalled to a width of 12 feet, had been made fit for motor traffic from Jandola, N.N.W. of Tank, through Sorarogha and Razmak to the Tochi valley at Isha.

This road runs through both Wazir and Mahsud territory. When this road was finished, the military occupation was nominally to cease, but actually the protection of the road and posts was to be taken over by 4,600 khassadars and 5,000 irregulars. Razmak and Jandola alone, together with the posts of the line of communications from Razmak to Isha, were to be permanently occupied by Regulars. The khassadars are local tribesmen who get 30 rupees a month and use their own rifles. The irregulars, indifferently called militia or scouts, wear our uniform and have British officers; two-thirds of them are recruited from cis-border and one-third from trans-border Pathans. So at least Lord Peel has said.

The circular road has been pushed on and is now practically finished. Some fine work has been done upon it by the companies of Sappers and Miners and the Pioneer battalions in a country which, for sternness and wildness, could give points to Caledonia. The Regular garrisons at Marobi and Piaza Raghza have been successfully withdrawn and replaced by khassadars. So, I believe, either has been or will be replaced our scout garrison at Wana, which has been withdrawn. The main garrison is at Razmak camp. Jandola is still held, as before. It is in the Batanni country, where the people are usually friendly. Sorarogha is also to be held. It is well in the Mahsud country, and is apparently to be garrisoned by irregulars. So are Kotkai, Sarwakai, and other posts.

It is to these inferior levies that the protection of our main posts is confided through a country which Lord Chelmsford accurately described the other day as 'a terrible welter of hills.' The country is still very disturbed, and the arrangements made for the passage of a traveller along the circular road read like operation orders. The garrisons are well wired in by night and 'shot up' by day. There were five brigades in Waziristan earlier in the year, besides an immense accumulation of other arms and services, exclusive of two other brigades at Bannu and Kohat, making seven in all out of twenty-two in India. How strong the future garrison is to be has not been announced, but it will probably be of four brigades, and nine battalions are down to move into Waziristan in the 1923-24 programmes of Indian Army reliefs. The aeroplanes are mainly at Dardoni, in the Tochi valley, and at Tank. They have been recently used against the Guri Khel, a sub-section of the Alizai clan, and in general have been kept very busy.

The invention of the half-forward policy appears to date from about July, 1922. It may be called a system of partial rather than of complete occupation. It is so important in itself, and may become so much more important in its consequences, that we must assume that it met with the approval not only of the Government of India, but with that of our General Staff and of the Cabinet, or a part of the Cabinet, at home. Mr. Lloyd George was Prime Minister at the time. Given the general situation to be as I have described it, both the policy and the strategy appear to combine all the mistakes that can be committed on this frontier, and are a deliberate reversal of the frontier policy which was formally announced in 1900 and 1902 by Lord Curzon to the tribes.

No sound reason has yet been given for the partial and apparently permanent military occupation of Waziristan. If it be militarily true that the control of Waziristan rests upon the control of the Mahsuds, and that the new arrangements promise such control, a policy based on this argument is a narrow one, and neglects the wider aspects of our problem. The tracks and valleys through this territory lead nowhere that we want to go. The garrison is and will be large and its supply costly, but it will only dominate the country within reach of its guns, and will not necessarily keep Wazir raiders out of India except in so far that it will attract an uncertain amount of hostility to itself.

It is affirmed that our future garrison will 'immobilise' 50,000 Mahsuds and Wazirs, but it is possibly more true to say that it will mobilise them. It is a standing provocation to what the tribesmen value more than life, namely, their independence, and will be a perpetual cause of strife and enmity, which may become exasperation owing to our air raids. There are reports that the Amir is much exercised because the Wazirs rose at his call for jehad in 1919, and he feels indebted to them. The other great tribes on the border are beginning to ask what it all means.

In Waziristan convoys, detachments, and individuals are liable to be constantly attacked, even if we are not told about such happenings, and our posts may become happy hunting-grounds for the young bloods of the tribes.

The new irregulars are much inferior to the Regulars, and their future causes anxiety. The moral effect of the beating which the tribesmen received will be gradually worn away, and the sums of money paid over to the Maliks and the tribal khassadars, as well as our promises of contracts, can only be regarded as blackmail. We certainly make allowances in other parts of the frontier to headmen and tribes for value received, but it is another thing to give them to Wazirs and Mahsuds who are hostile, and will only use our money to buy more and better arms. We need an explanation why it is considered legitimate to embark upon this great commitment in Waziristan, to which no end can be seen, when the Indian Army, and the British troops in India besides, are suffering from the effect of the Inchcape axe and the general financial stringency.

The Indian Budget of 1922-23, provided 2·13 crores for Waziristan, but the actual sums expended were 3·85 crores. Worst of all, from a military point of view, is the chaining down of such important bodies of troops for an indefinite period of time, and the corresponding diminution of the much-attenuated military forces in India available for general purposes. We could understand the increase of the Army to meet a great and new commitment, but to accept this commitment and to reduce the Army is an absurdity.

Training must suffer, and the discontent of troops without the amenities of cantonments and bazaars, enclosed by barbed wire, and quartered in winter at Razmak camp, at an elevation of 7,000 feet, will be considerable. It is necessary for everyone who studies the subject to regard it not only from the local military point of view, but also from that of the Wazirs and the Afghans. Since a Whig Government at home sent Lord Auckland to India in 1836 instead of Lord Heytesbury, who had already been appointed by Sir Robert Peel, our policy in Afghanistan has made a series of mistakes of a very costly kind, and we should do well, as we regard all this matter, to place ourselves in the position of the reigning Amir and of the tribes as a whole.

A Clear Policy Needed

It must be premised that no one, so far as I am aware, desires to make any change in the arrangements in the Khyber, the Bolan, and at Quetta. The two great lines of communication which lead respectively to Kabul and Kandahar are already, and will increasingly become, international highways, and must always remain in the firm grip of our Regular British and Indian Forces. We cannot permit a repetition of the great humiliations which we have suffered in the Khyber. It is the most famous continental gateway of India, and the pass which many great conquerors have trodden. The new railway through the pass is now completed to Landi Khana. Quetta and the Bolan to New Chaman, are also vital, and there can be no suggestion of committing their guardianship to any hands but those of the Army in India. Similarly, there is no reason to interfere in the Baluchistan or with our defensive arrangements there, whether in British Baluchistan, the Agency Territories, or the Native States, even though the murder of Major Finnis shows that the Zhob is not as well protected as it should be. Whether Sir Robert Sandeman's system could ever have been applied to Wazirs and Afridis one may doubt, but that it applied admirably to the special conditions of Baluchistan few have ever questioned.

The real question is whether there is a means for protecting the rest of our border better than that now being applied in Waziristan. The first thing to decide, before giving our answer, is what we want there. Do we, or do we not, intend to maintain the virtual independence of the tribes? Do we, or do we not, propose to extend our administration over the tribes up to the Durand line? Do we, or do we not, intend to introduce generally this half-forward system upon which we have embarked in Waziristan? Have we measured, and do we accept, for good reasons, the consequences of the half-forward policy? What are these reasons?

There is a good case, in right and logic, for occupying the whole borderland up to our political frontier with Afghanistan, but in the event of our deciding upon this policy it will now mean a great military effort, which will demand heavy sacrifices and will tie up in the tribal territory for long the greatest part of the Army in India. Neither the great changes which I have recorded, nor these troubled times, nor the financial and military sacrifices involved, recommend such proceedings now, while as for the half-forward policy now in operation in Waziristan, it appears to combine the defects of all other systems and the advantages of none of them.

No system of border protection is perfect, nor, because a system succeeds in one part of the border, does it necessarily apply to another. So long as the tribes delight in looting, murdering, and fighting, they will get what they want, and so long as a wave of fanatical religious feeling, whether spontaneous or stirred up by an enemy, is liable to sweep along the frontier, so long are we always exposed to partial or general outbreaks. No system can guarantee that the border will be absolutely safe, nor that expeditions, whether on a small scale or a large, will not become necessary. It is not a question of counsels of perfection, but of doing the best we can to secure orderly government on the border, and to improve, and not harm, our relations with the tribes, and through the tribes with the Amir.

If we are not prepared to apply the forward policy, and if the half-forward policy becomes demonstrably a cause of dispersion, weakness, disorder, and political instability, we should do well to accept tribal independence definitely, to reduce our trans-border commitments, which irritate the tribes so much, and to set about the business of making the tribal borderland more secure than it is now. The Amir certainly has an influence over the tribes, and his desire to make them a cactus hedge between himself and us is quite comprehensible. We cannot very easily apply a general principle without making exceptions, for, while some may wish to withdraw our garrison from Chitral, only a minority would approve that we should abandon the Malakand (where the people of Swat wish us to remain), the Kurram Valley (where the Turis have always been faithful to us), the Tochi, possibly as far as Miranshah (Dardoni), and the Thal-Idak lateral line.

It may be said that a section of the Wazirs invited us to Razmak, but they did so to stop the encroachments of the Mahsuds, and it is not our business to interfere in the domestic politics of the tribes. In general, a correct principle to follow is to reduce trans-border commitments to the minimum, sure as we ought to be that they are the most fruitful source of irritation to the tribes and the Amir, a mark for all the restless elements to shoot at, and a most disagreeable encumbrance when their relief or reinforcement becomes necessary. It is better by far to be able to hold together the very meagre military forces we have in our Indian cantonments, where we can regularly train them, keep them healthy and make them contented, than to scatter them in isolated garrisons like Razmak, where they may impose upon us a relief expedition when we have more important business in hand elsewhere.

The Wazirs, fighting in their own wild hill country against the invader, are formidable foes. Outside their own country they count for little. It is not, perhaps, a mere coincidence that, while Waziristan, where we have had trans-border commitments for long, rose at once in 1919 to the call of the Amir for jehad, the Afridis and the Mohmands,in whose territories we have no commitments, actually took our side. The Afridis pillaged Jalalabad while the Mohmands looted Dacca, and turned back an Afghan lashkar when it sought to enter their territory. On the other hand, the Wazirs and Mahsuds captured our posts at Wana and Datta Khel, and with them took some 1,500 rifles and at least one and a half million cartridges. Not that this is of supreme importance, as these gentry are already armed to the teeth, are in constant touch with the Afghans, and receive ammunition from Kabul.

We no doubt gained much in 1897 and 1920 by proving to the Afridis and Wazirs that their most inaccessible fastnesses are not secure from our arms ; but, whereas our withdrawal from Tirah and our subsequent conduct on this part of the frontier have almost convinced the Afridis and Orakzais that we do not covet their territories, we have mistakenly proved the reverse to the Wazirs, and our unfortunate action in this case may cause doubts about our intentions to circulate along the frontier and at Kabul. We must not be surprised if it does. We are asking for it.

A Protected Border

Now necessarily, if we restrict ourselves to a protected border, we shall be told that we are standing on the defensive and that the offensive is the soul of war. So it is, when war comes. But in peace a defensive attitude is the rule. When Hadrian built his wall from the Solway Firth to the Tyne he adopted the defensive on his Caledonian front because Caledonia then, like our tribal borderland in India now, offered nothing worth having. So he built his wall and his ditch, and particularly made sure that a good lateral road should run behind his wall from end to end. He made cohort camps every four miles, and he placed watch-towers along his road. It was the 'close border' of the Romans and it served. This was an exception to the unhappy conduct of Roman frontier defence to which I have already alluded.

The 'close border' is an old term in India. It has never been applied there, nor could be with the old technical equipment, while the name points to blockade, which is the last thing we want on our tribal border. Therefore we must reject the term because it is what we do not want. We like the trans-border Pathans on the whole. We understand them, and they us. We have many relations with them, and they have many relations with India. They are born traders, and we find them ranging all over India and as far as the Bengal bazaars. Our object is to extend these relations and to encourage confidence and trans-border trade. It is not a close border, but an open and protected border, that we need.

A protected border in times of peace is a police affair. In the North-West Frontier Province we have a small force, not over 2,400 strong, called the Frontier Constabulary, meanly found in many respects, but whose remarkable successes of late in preventing or avenging robberies and murders by the Pathans across the border have greatly interested all who keep an eye upon events there. It is locally recruited, and is commanded by British officers recruited from the now very efficient Indian police for a term of years. There is a commandant of the corps, also a police officer, at the headquarters of the civil Government, who makes frequent tours of inspection. There are Indian officers as well as British, and all ranks receive a higher rate of pay than the ordinary police. It is a kind of frontier police militia, and the officering and training by police officers is the main distinction between it and the ordinary frontier militia, which is also under the civil Government, and serves across the frontier in the agencies such as Kurram, Tochi, etc., and has Army officers over it.

All these militia, scouts, khassadars, and what not are very useful in their way, so long as we do not any more regard them as a first line of military defence, and give them charge of the keys of our front doors or fail to support them by Regulars at need. It was this error which discredited the Curzon militia scheme for a time, but the Curzon policy was sound.

The present system of border protection is divided up into civil districts. Each deputy commissioner sets his own machinery in motion, and his district officer and constabulary are under him. Each deputy commissioner can call upon the nearest Regulars for assistance when he thinks it necessary, and the Regulars—British or Indian—respond with alacrity when there is a chance of a scrap. This system works very smoothly, and the deputy commissioner can on his own authority take all the necessary measures to deal with a raiding gang, except that reprisals across the border must be sanctioned by the Chief Commissioner. It is really remarkable how successful this system has been, considering the length of the frontier and the small numbers and means of the constabulary. Were it not for the new policy in Waziristan, we should decide that things had much improved during the last three years.

Factors of Success

But success depends upon certain factors which must be understood. First a good system of lateral roads like Hadrian's is absolutely indispensable to enable motor transport to exploit the advantages of rapidity and surprise. The constabulary are weak, but very mobile. It is the roads and the motor transport which give mobility. There are such roads in the Peshawar district, as many a snipe-shooting subaltern knows, and the greatest successes of the constabulary have occurred in this district, although it is the most exposed point on the whole frontier and half-encircled by warlike tribes. But in the province generally, and particularly on the frontiers of Waziristan on the Derajat side, there are few or none of such roads, and until they are made the protected frontier cannot exist. Telephones and aeroplanes are also an indispensable part of the equipment, and if wireless could be procured, it would be an excellent thing. The best thing in Mr. Bray's speech of March 1923 was the proposal to construct 100 miles of road on the Derajat border, but a lateral road, without the rest of the organisation which I have described, has no defensive virtues.

The people who live on our side of the border are a good set of folk, and their khans are men of whom not enough has been made in the past. These people are a virile lot, and there are many old soldiers among them. They are ready and willing to defend themselves. They work in very cordially with the constabulary and the ordinary armed police when raiders are about. The chighas, or pursuit parties, raise a hue and cry and are quite ready to fight when they are decently armed. But they are poorly armed at present, and the defensive organisation of our frontier villages still leaves much to be desired. Lateral roads, motor transport, telephones, wireless, more constabulary, and a proper organisation of the frontier villages and of the hue and cry, all are indispensable parts of a good protected frontier scheme, but hitherto the resources allotted to these important objects and services have been absurdly inadequate.

If the present constabulary were doubled as a beginning, the necessary technical equipment and arms supplied, a good air force made available for observation and reprisals, and 100 lakhs devoted to the construction of lateral roads along our border, where they would serve trade as well as defence, the protected border would be a fact, and we could then withdraw our expensive garrisons from Waziristan, and with the greater pleasure, because they do not protect our border from roads.

Whether the Indian reformers in the Legislative Assembly will take sufficient interest in their fellow-citizens on the border to urge this reform when they meet again, and whether they will vote the money for it, remain to be seen. The border Pathan is notoriously the most expert thief and blood-thirsty raider in the world. We have had too many recent examples of his ruthlessness to underrate him.

One particularly lucrative line of his business of recent years has been kidnapping, and it would be interesting to know how many of our Indian fellow-subjects have been kidnapped during the last ten years, and how many scores of thousands of rupees have been paid by their relatives for ransom. The reduction in the numbers of raids and murders during the years 1921 to 1923 has been gratifying, but outrages cannot be stopped until the protective system which I have described is fully fledged and put into general application. They can be, and are being, met to some extent by the prompt seizure of trans-border Pathans in India who are believed to be related to the raiding gangs.

The firm hand of the High Commissioner, Sir John Maffey, in this matter has been much appreciated. It is better to protect our own border from raiders than to go whining to the Amir when the ruffians slip across, and then take refuge in his territory after raiding ours.

It is, perhaps, of secondary importance, but many people consider that a restoration of some equivalent of the old 'Piffers'—the very popular and efficient old Punjab Frontier Force—is desirable. This would, to some extent, be only a paper transaction, as the locations and garrisons would not be affected, but a frontier force permanently located in the North-West Frontier Province would be popular as well as useful. The frontier needs, and normally has, a Regular garrison of seven infantry brigades at Peshawar, Kohat, Bannu, Dera Ismail Khan, and Abbottabad, as a backing to what should be the protected frontier in peace time, and it is better that it should be composed of regiments trained and accustomed to the frontier than of a succession of green troops, who have to buy their experience in turn. Many Piffer regiments would not allow police in their cantonments. They knew the frontier and could look after themselves.

But it must be observed that when we chance to be engaged on a large scale upon operations in Afghanistan we must make up our minds that we shall still need Regular troops as a backing to the civil constabulary between Quetta and Peshawar, and it is the business of the Mobilisation Branch at Army Headquarters in India to allot them, after consulting the civil Government, remembering that it is on these occasions that the border is most restless. When Turkey is on the war-path, and for a year or so afterwards, the restlessness is most marked. The complement of good lateral roads along our border is an equally good lateral system of railways between Quetta and Peshawar, to serve the Army in India when it is called in.

It must, however, be understood that though the trans-border Pathan is a fine fellow on his native hillside, he is out of his element in the plains. So far as I can recall, the largest raiding party that has ever crossed our border has not numbered more than 600 men, and a lashkar that embarked on making war upon us in India would disperse as quickly as snow in summer. Ten thousand tribesmen have before now approached the border, as at Shabkadar, but have not crossed it, because they were well beaten.

It is the smaller bands which cause most trouble and raid down to the Indus and over it. Usually the numbers of the raiders are small, and it does not, as a rule, need more than constabulary, increased in numbers and organised as I have described, to deal with them. To meet large agglomerations we want the Piffers, and the Guides as they used to be.

The Air Force is a valuable arm for observation and reprisals, and it may, on this border, be sometimes a substitute for other forces. The trans-border Pathans know what baramta means. But public opinion here will not much longer tolerate the bombing of villages as a police operation or as an act of war unless adequate notice is given for the women and children to be placed in safety.

A Searchlight Needed

It is necessary to throw a searchlight upon Indian frontier policy, because it is obscure and its implications are not at present understood in England. We have not been properly informed upon the objects of the Government of India, and Lord Montagu of Beaulieu's request, on May 3, 1923, for papers from the General Staff in India has been ignored. For all that the public have been allowed to know about the opinions of the General Staff since the war, this organ of the Higher Command might not exist.

The new system which is being applied in Waziristan seems in every respect detrimental to our interests and highly irritating to the tribes and the Amir. It cuts directly across, and is absolutely antagonistic to, the assurances given by Lord Curzon to the tribes in his speeches in full Durbar at Quetta on April 12, 1900, and at Peshawar on April 26, 1902, speeches inspired by a wise and correct comprehension of our true frontier policy. It is not a partial and indefinitely prolonged military occupation of Waziristan that we need to protect our frontier. It is the series of civil reforms which I have enumerated, and if they were rapidly put through they would give the frontier, in eighteen months or two years, far better protection than this Waziristan entanglement, which gives little or no direct protection to our border, threatens to create endless trouble, violates our old assurances to the tribes, and has really nothing to recommend it. Let no one suppose that a weakening of our garrisons in Waziristan, such as is in progress now, will serve. It is merely an inadmissible gamble. They must come out.

No doubt it is highly inexpedient to change unnecessarily a policy deliberately embarked upon, but if a policy is radically unsound and dangerous there is no reason for persisting in it. It is very desirable that a new Government in London should re-examine a plan which they inherited and did not initiate. They should call to their councils at the Defence Committee the best Anglo-Indian civil and military advice available before they finally commit themselves and us to a costly, foolish, and hazardous enterprise which, if persisted in, may lead to a recurrence of the events of 1897, if to nothing worse.

There is no doubt at all that the great rising of 1897 took place because the suspicions of the tribes had been excited by the extension of British influence and the establishment of British garrisons in what had formerly been independent territory, and that the mullahs, and exaggerated rumours countenanced by the Afghan authorities, fanned the smouldering fires into a flame. From the Tochi to the Malakand our outposts were suddenly and simultaneously attacked. Have our authorities forgotten that costly lesson ? I presume that they must have done so, because in the debate on Waziristan in the House of Lords on May 3, 1923, not one word was said of the political effect of our Waziristan policy upon the Amir or the tribes.

Finally I must come back to a point on which I have already laid stress. Success in India and on the frontier depends mainly upon the selection of the men whom we send there, and upon the wise policy of the home Government. Lord Heytesbury was totally opposed to the policy of Lord Auckland, who forced on the first Afghan War, and, except Sir John Hobhouse, every great Indian statesman of experience, including Lord William Bentinck, the Duke of Wellington, Lord Wellesley and Mr. Elphinstone, agreed with Heytesbury. So did most of the reflecting minds in India and England at the time.

Our policy to-day is the old Auckland policy in another form, and we must kill it before it ruins not only our relations with the tribes but with the Amir. The Government must find time to examine the whole question and to lay down that our policy must be one of friendship and neighbourly relations with Afghanistan and the tribes. I do not think that we should wait for the spring and summer.

When our Cabinet mullahs become Talib-ul-ilm we shall get on. I believe that there are very few civil or political officers on the frontier who agree with this half-forward policy, and to enlighten us we need to have before us the opinion of the High Commissioner, the Deputy Commissioners, and the political officers who are in constant touch with and control of our border relations and know what they are talking about. Naturally we also desire to have the views of the General Staff in India and at home, and the reasons why Mr. Lloyd George's Government, some time between July and September, 1922, sanctioned the partial and permanent occupation of Waziristan.

Our success on the frontier has been due in the past to the high character of the men who we employed there—men like the Lawrences, Mackeson, Herbert Edwardes, James Abbott, John Nicholson, Lumsden, Jacob, Sandeman, and their school. It is men like these, men who are men, of high character, great gentlemen, and of conciliatory disposition but firm texture of mind, who control by their personality the wild border which dearly loves a man and will follow him anywhere. It was these men who made the border a source of strength and not of weakness in the Mutiny days.

Have we forgotten that John Lawrence and Herbert Edwardes had made the Amir a friend before that event, that all the wild tribes of the frontier flocked in to fill up the gaps in the native army at Peshawar when it had been disarmed, that hundreds of Khans and Urbabs appeared and were profuse of offers of service, and that all the chiefs of the Derajat instantly took up arms, raised horse and foot, and hurried to our aid? It was by the wise policy of the great men of the frontier that these things came about, and the same thing will happen again in time of need if we follow the precedents of the past. That there are such men on the border to-day is known, and it is better to have a man on the frontier than an army, if only the policy of the Government is sane.

THE FORWARD SCHOOL

It was apparently not possible for Sandemanians to take up the challenge given in the previous pages. I should have been much better satisfied had they done so. But it is the policy of the last four years that is in question, and those who know the story from the inside are few in number, and all in official positions, which prevent them from replying, however anxious they may be to do so. In former days a long series of Blue Books enabled the public to keep touch with events, but since 1919 there has been a curious reluctance, to use no more forcible term, to publish anything of serious value. In these circumstances, the best course seems to be to endeavour to state the forward case myself by the help of conversations and correspondence since my previous pages appeared, and I hope to do so fairly because I regard the whole question as a fascinating problem of Empire which we, our fathers, and our grandfathers have found as insoluble as the problem of the independent tribes on the mountain frontier of Italy was found by Rome till the days of Augustus.

POLICY AND GENERAL STAFFS

Some recent articles published in England have described the policy now in course of application in Waziristan as " General Staff policy." This is obviously inaccurate and creates a prejudice. A General Staff has no policy. Policy is the province of government. In India, policy is decided by the Governor-General in Council, and if this policy be submitted to, and approved by, the Home Government, then it becomes British

Government policy as well. A General Staff's business is to propose the military measures needed to carry out a political decision, but it has no political decisions to make on its own initiative and account.

It is, of course, true that India is, and always has been, largely permeated by military opinion, and that soldiers have a much greater influence upon affairs at Simla and Delhi than they have in London. It is also true that the General Staff in India, by their continuous and energetic support of the policy laid down four years ago, have been largely instrumental in preventing it from being dropped or mutilated beyond recognition. When the question was referred to London in 1922, the General Staff had a strong team to represent them, including competent and experienced officers well acquainted with the whole position. No one can blame them for having supported the policy of the Government of India with all their might. They had learnt in France to dig in and consolidate a position won. This they did.

The forward policy in Waziristan was decided upon by Lord Chelmsford and his Council in March, 1920, when Sir Charles Monro was Commander-in-Chief and a member of the Government of India. When the policy was settled the General Staff put up the plans to carry it out. Lord Rawlinson approved of the plans when he succeeded Sir Charles Monro in November, 1920, and carried them out so far as finance permitted. It was in 1922 that the case was examined in London by the Defence Committee and the plans in general approved by them, but this decision was not communicated at the time to all the members of the Cabinet. We were then, of course, under Coalition rule.

Waziristan

Why was the policy of the partial and permanent occupation of Waziristan decided upon by the Government of India and approved by the Defence Committee in London? We are told it was because the Wazirs and the Mahsuds had been a thorn in our flesh for sixty years; because we had been compelled to initiate operations against them upon seventeen different occasions; because they were the most incorrigible raiders and bandits on the border; because the number of their victims and the extent of their raids and depredations had made decisive action imperative; because they had made war upon us when the Amir instigated them to jehad in 1919; and because occupation was the only course left when every other form of provest and pressure had failed.

The "close-border" policy had been tried for twenty years since Lord George Hamilton's famous despatch to Lord Curzon in 1898, which laid down the principles of that policy. It is claimed that it did not bring peace to our border districts contiguous to Waziristan, but produced the events of 1900-02, 1907, and 1917, and finally failed in the crucial test of 1919. Every year, we are told, the tribes are becoming a greater menace to the peace of the border on account of their improved armament and better facilities for obtaining arms and ammunition. The Waziristan campaign of 1919-21 required the employment of 45,000 men. If another campaign has to be undertaken it is suggested that the numbers required will be proportionately greater. The present policy is stated to be the only one which shows signs of producing such a situation in the future that the number of troops permanently located on the frontier may be capable of reduction.

The military plan was to get behind the tribes and to control them by garrisons and the extension of motor roads. The plan aimed more at the military control of the tribes from within than at the protection of our border from without, and the Air Force was constantly employed. Have these plans effected their purpose? It is too early to say for certain, but the number of raids on our frontier has fallen from 391 in 1919-20 to 194 in 1921-22, and to 131 in 1922-23. These figures do not include the attacks upon posts, convoys, and individuals within Waziristan itself, which have also shown a marked decrease. The progress noticed is partly due, as declared by the Report on the Administration of the Border of the N.W. Frontier Province for 1922-23, to the energy displayed by the district and police officers, the frontier constabulary police, the Political Agents, and the village chighas, within our own administered territory. For financial reasons a full forward policy has not been applied, and it is hoped that as regards actual military occupation it may never be necessary. What has now been done is regarded by the General Staff as a step in the direction of obtaining full political control. Whether our cantonments at Razmak will achieve this only time can show, but some extension of road construction will certainly be necessary.

Road Construction

For example, Wana, the scene of one of the deplorable Militia disasters of 1919, is outside the present web of control, and the south-west part of Waziristan generally is inadequately watched. A motor road to connect Razmak with Wana, via Ladha and the Lare Lar Pass, is most desirable, or, failing this, an extension to Wana of the Jandola-Sarwekai road now under construction. An u timate continuation to Fort Sandeman, via Kajuri Kach and Moghul Kot, so linking up the Zhob with Waziristan, would strengthen our position greatly. We should then have roads at the back of nearly all the most troublesome sections of the Mahsud and Darwesh Khel Wazirs, but at present financial reasons have prevented these roads from being made, and consequently the full road scheme which the General Staff advocate is not yet being carried out in its entirety.

Ultimately, the desire is to have a good motor road to connect Peshawar and Quetta, via Kohat, Thal, Spinwam, Idak, Razmak, Wana, Kajuri Kach, and Fort Sandeman. The two large garrisons at the ends of the line can then, it is thought, be tapped in case of trouble, with a very quieting effect on the border, and a possible reduction of the frontier garrisons. It is hoped that the narrow-gauge line from Hindubagh to Fort Sandeman will shortly be commenced, and if at some future date we get a railway from Peshawar to Quetta we are told that we shall stand in a favourable position. That seems to be the General Staff idea.

Concerning the garrisons, it is known that before the war there stood two independent mixed brigades in Waziristan—at Bannu and Dera Ismail Khan respectively—to guard the frontier and support the Militia posts. At present there are four brigades, in addition to the Scouts and Khassadars, owing to the improved armament of the tribes and the general conditions existing in Waziristan since the Afghan incursions of 1919. It is alleged that the same Regular garrison will be required outside Waziristan if we retire to the Tochi and to Jandola. The places which should be held in case of a retirement are said to be Dardoni, Miranshah, Idak, Kajuri, Saidgi, and Bannu in the Tochi, of which the first and last will require at least two battalions each. In the Derajat there would have to be held Jandola, Khirgi, Manzai, Kaur Bridge, Tank, Dera Ismail Khan, and Kalabagh, exclusive of any new posts which might be needed. The General Staff in India, it need scarcely be said, offer the strongest possible opposition, on military grounds, to any such retirement. They point, moreover, to the healthiness of the Razmak site for British and Gurkha troops in comparison with the unpopularity and unhealthiness of such stifling towns as Tank, Dera Ismail Khan, and other garrisons of the Indus Valley.

The General Question

The rival schools have no great differences of opinion concerning a large part of the frontier. Though there are a few Lawrentians who would not hold Chitral, they are in a minority of their own party, and there are few dissentients form the opinion that, north of the Kabul river, there is no need to impose our administration so long as there are rulers like the Mehtar of Chitral, the Miangul of Swat, and the Nawab of Dir to exercise authority and remain friends with us. Similarly, in Baluchistan and the Zhob, neither school desires to alter the situation. Here, apart from Wazir raids into the Zhob, which would cease if we controlled Waziristan, we are more or less at peace. Our administrative and political frontiers coincide, and no one wishes to alter anything. It is also almost common ground that the Khyber and Chaman must remain in the hands of Regular troops. There is certainly a difficult bit of the border where the Mohmands and Shinwaris live, for both are partly on our side and partly on the Afghan side of the frontier. This cannot be altered unless we advance to the Kunar river, which is admitted by the forward school to be neither desirable nor practical politics. No one that I know of proposes to alter the frontier of Afghanistan.

The battle-ground of the schools is the frontier from the Kabul river to the Gomal. I have now given the views of both schools about Waziristan, and it remains to define the position of the forward school in the country of the Afridis and Orakzais. Although the extension of our political control, it is hoped, will continue gradually to become more effective, as it is claimed to have become in the last few years, no one desires to interfere with these people so long as they remain quiet and behave decently. But, should 1897 recur, and operations in Tirah be forced upon us, the whole question of road construction and the location of the troops now in the Khyber, Kohat, and the Kurram would, in the opinion of the forward school, have to be reconsidered.

Naturally the forward school, like their opponents, quote from a long list of the statesmen and warriors who have advocated their policy. The forward school admit that many of the circumstances have changed since some of these authorities spoke and particularly that, for the present, the idea of fighting for Afghanistan against Russia need not be considered; but they maintain that the broad principles of the policy of the older statesmen and soldiers hold good, and are of even greater weight under present conditions.

Effects on Afghanistan

As regards the effect of our frontier policy upon our relations with the Amir, the forward school maintain that with Afghanistan now a completely independent State, and with its rapid development towards civilisation under the impulse and guidance of the Amir, our old position of ineffective control and shirked responsibility on the British side of the Durand line is an anachronism, and conducive only to difficulties and misunderstandings with Afghanistan. Such conditions, they urge, may cause the Amir some inward satisfaction, as leaving a wider field open to Afghan ambitions, which are, perhaps, as yet hardly formulated. But they add that our policy on the British side of the frontier must be dictated by our responsibilities and requirements, and that Afghanistan can have no more to say in the matter than we would presume to claim to have in Afghan policy on the Afghan side of the frontier. The forward school maintain with vigour that a clear recognition of this will form the best assurance of mutual respect and neighbourly relations between ourselves and Afghanistan. In the case of the tribes they propose to apply the maxim, "Do as you would be done by." So long as a tribe behaves well it will be treated well and not interfered with, but if it continually misbehaves, intrigues with Afghans, Bolsheviks, or Indian seditionists, and bedevils the border, then it will be brought to book, but no longer by the process of "raid and scuttle." The forward school are not for such temporary expedients, and the principle on which they desire to act is that, when action is forced upon us, our remedial measures should have some prospect of permanency. The logical development of such a policy implies ultimate control up to the Durand line.

One common belief unites both schools, namely, that the unrest on the border, in so far as it is not due to natural lawlessness, fanaticism, or instigation from outside, is due to poverty and absence of means for improving trade, and the sparse cultivation and afforestation of the mountain valleys and hills. The tribesmen, in fact, have not the means to earn a livelihood decently. Whatever policy is eventually adopted, this crying need of the border must be met, and I believe that Lord Olivier has already realised it, and has a strong desire to take the matter up.

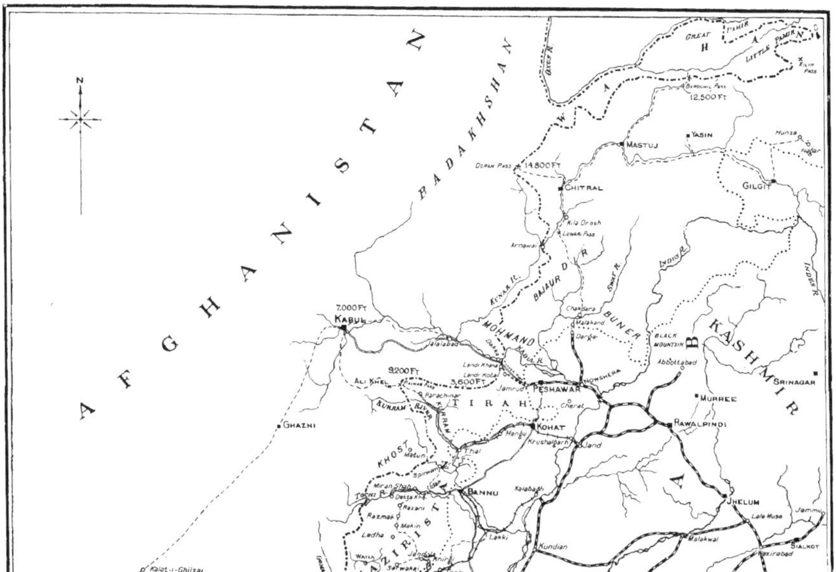

THE NORTH-WEST FRONTIER OF INDIA.

p. 278.

APPENDIX I.

RECOMMENDATIONS

OF THE

NATIONAL AND IMPERIAL DEFENCE COMMITTEE

AS APPROVED BY HIS MAJESTY'S GOVERNMENT,

UPON

I.—The Relations of the Navy and the Air Force.

II.—The Co-ordination of the Defence Forces.

Cmd. 1938. 1923.

APPENDIX I

I.—THE RELATIONS OF THE NAVY AND THE AIR FORCE.

The Main Committee observe that the Special Sub-Committee are not altogether satisfied with the term 'Seconded' which they have used in describing the status of Naval officers serving in the Fleet Air Arm. It is clear that the actual position of a Naval officer seconded to the Air Service afloat differs from a seconded officer as hitherto understood. For example, he continues to be under the ultimate tactical and disciplinary orders at sea of an officer of the Service to which he permanently belongs, whilst he is serving under the immediate command of an officer of the Service to which he is temporarily attached. Moreover, the work he is called on to do in his temporary Service directly relates to the needs of his parent Service, and he carries it out under the eyes of a commanding officer of his parent Service. The Main Committee therefore consider the term 'Seconded' not altogether appropriate. They would prefer the more general term 'Attached,' provided the conditions of attachment necessary in this particular case are clearly understood, as follows: These Naval officers belonging to the Fleet Air Arm are therefore to be attached to the Air Service. But whilst after their training under the Air Service they will be posted to units of the Navy by the Air Service on the nomination of the Admiralty and will serve under the immediate command of Air Service officers, yet afloat they will be under the ultimate command of the Naval officers of the Fleet. Though their temporary pay and their acting rank will be determined by the Air Service, after consultation with the Admiralty, yet their permanent rank and promotion will continue to be subject to the regulations of the Navy. Lastly, though they must serve with the Air Service and not be moved from it during the period of their attachment without the consent of the Air Ministry, yet as regards discipline and status afloat they will in all respects be in precisely the same position as officers of the Royal Marines.

With regard to uniform, the officers and men generally of the Fleet Air Arm will wear a distinguishing badge on their uniform. If the Admiralty desire, the Naval officers attached to the Fleet Air Arm will retain their Naval uniform with the same badge.

If any Naval other ranks and ratings are employed in the Fleet Air Arm, the same principle as regards uniform and badges would apply.

REPORT OF THE SPECIAL SUB-COMMITTEE

PART I.

The General Problem

1. Your Sub-Committee were instructed to inquire into the relations of the Navy and the Air Force as regards the control of Fleet air work, and on this point they have taken a great deal of evidence from witnesses representing both Departments.

In addition to this, two of our members, Lord Peel and Lord Weir, accompanied by Sir Maurice Hankey (Secretary), paid a visit to the aerodromes in the neighbourhood of Portsmouth and the Aircraft Carriers *Argus* and *Eagle*.

2. In the course of the somewhat acute controversy between the two Departments, a very large number of points have been raised whose importance is subsidiary and illustrative. This was inevitable, and probably desirable; but the broad principles lying at the root of the policies respectively advocated by the contending parties are perfectly intelligible, quite apart from the detailed arguments and counter arguments by which they are supported.

3. To the Air Ministry was given in 1917 control over the air forces of the country. The Department was at one stroke placed on a par with the Admiralty and the War Office, for, though aerial warfare is a growth of yesterday, it was felt that its rapidly increasing importance required an organisation which would ensure its development under the most favourable conditions. In these circumstances it was only natural that the Air Ministry should strongly object to being partially dismembered so soon after it had been brought into existence, the more so as in the early days of this dispute the severance of the Fleet Air Arm from the rest of the Air Force meant an immensely larger proportionate loss than it would mean at the present moment. But though the injury involved in the complete removal from their jurisdiction of the Fleet Air Arm is relatively diminished, it remains in principle the same, and if carried out to its logical conclusions would, in the opinion of the Department, lead to many administrative difficulties and much overlapping, while in the region of supply and research it would hamper progress and increase expense.

4. The Admiralty case is not less worthy of most serious consideration. In their view a Fleet Air Arm is now as necessary to a Fleet as cruisers, destroyers or submarines. Aerial reconnaissance and aerial 'spotting' are as strictly naval operations as gunnery, torpedo work and wireless telegraphy. It seems to them intolerable that, while they are responsible for the safety and success of our Battle Fleets, the air work on which that safety and that success in large measure depend should be performed by persons belonging to another Service, imbued by different traditions, and looking for support and promotion to a different Department.

5. Your Sub-Committee have the greatest sympathy with both these points of view, and they have given much anxious thought to the question of how they can best be dealt with. They do not think that the present system can remain altogether unchanged; neither do they think it possible to sever completely the Air organisation which does work for the Fleet from the Air organisation which is responsible for Home Defence against

air attack and for co-operating with the Army in other theatres of operation. A course somewhere between these two extremes is the one we recommend, but before describing our suggestions in detail it seems desirable to give a brief sketch of the system now in course of development along the line where the two Services come into contact.

PART II

THE PRESENT SYSTEM

The General System.

6. The general system is that the Air Ministry raises, trains and maintains the Fleet Air Arm. At sea the Fleet Air Arm comes under the operational and disciplinary control of the Admiralty, which designs, builds and maintains the carriers.

The functions of the Fleet Air Arm comprise aerial reconnaissance, naval gunnery spotting, bombing and fighting.

The Fleet Air Arm operates at sea from a carrier, which is a floating aerodrome. In addition, in time of war, capital ships and cruisers will embark aeroplanes, but at present the risk to personnel and *matériel* is too great to justify this in time of peace.

Air Force squadrons also operate over sea from shore bases in co-operation with the Navy, but these are not at present in question.

The numbers of the Royal Air Force personnel at present employed with the Fleet Air Arm are 115 officers and 681 other ranks.

Policy.

7. Naval Air policy is concerted by the Air Staff and Admiralty War Staff. Initiative may come from either side, but Naval officers of high rank do not, under the present system, have many opportunities of obtaining experience or training in air matters to fit them for this work.

The Coastal Area.

8. Apart from matters of policy, *liaison* is secured between the Navy and the Air Force by a special machinery termed the 'Coastal Area' Organisation. This is provided by the Air Ministry for the purpose of administering the Fleet Air Arm on land, subject to the general control of the Air Ministry. The Coastal Area is commanded by a senior Air Force Officer (the present occupant is an ex-naval officer) who is responsible for the maintenance at full strength of the personnel and *matériel* of the Fleet Air Arm, and for its special training for naval purposes. He is the adviser to the Admiralty on Air matters other than policy, and communicates directly with the Commander-in-Chief, Atlantic Fleet, on all matters connected with the Fleet's requirements.

9. Under the Air Officer Commanding the Coastal Area are two Group Captains, with headquarters at Leuchars (the Forth) and Lee-on-Solent respectively. In the case of larger operations involving the embarkation of aircraft or the carrying out of operations from shore bases, the Naval Commander-in-Chief notifies his requirements direct to the Air Officer Commanding Coastal Area. For minor operations he deals directly with one or other of the Group Captains.

10. Corresponding arrangements are made at Malta for the Mediterranean Fleet.

Position on Board Carriers.

11. As soon as the Air Units are on board the carriers they come under the orders of the naval Commander-in-Chief, and are altogether outside the control of the Coastal Area Organisation, which is then, as we have already noted, only responsible for keeping units up to strength in air personnel and *matériel*. The orders for flying are given by the naval Commander-in-Chief (who has an Air Force Officer on his Staff) to the captain of the Aircraft Carrier, who is, of course, a naval officer, and he gives his orders to the senior Air Force Officer on board. The Royal Air Force personnel are responsible for the cleanliness and maintenance of their own quarters and some other parts of the ship, such as the aeroplane deck. They can be employed on the general duties of the ship. Difficulties have, however, been made in regard to the performance of particular duties, such as the training and use of Air Force personnel for passing ammunition to the guns.

12. A certain amount of co-operation is practised between the skilled mechanics of both Services. The Royal Air Force may, for example, utilise the services of the ship's blacksmith for certain repairs to aeroplane engines, and the Royal Air Force may undertake repairs to motor-boat engines. Such co-operation is naturally limited by the highly specialised training which is inevitable in each Service.

Supply of Personnel.

13. The Air Ministry are responsible for the supply of officers and other ranks for duty with the Fleet Air Arm, with the exception of certain naval officers, who, after training in Air Force establishments, are employed as observers in 'spotting' machines and some naval ratings employed as wireless telegraph operators.

14. It is of importance to note that the Royal Air Force possesses a number of officers who were formerly in the Royal Naval Air Service, many of whom were trained as naval officers. The senior positions in the Fleet Air Arm are at present largely filled by these officers. The original intention of the Air Ministry was that, in order to replace these, some 30 per cent. of the officers of the Fleet Air Arm should consist of naval officers seconded for a term of years. Unfortunately, however, only a very few naval officers responded to the invitation to volunteer for this service.

15. It may be worth noting that the general system for the supply of Air Force officers to meet future demands is divided into two classes:

(*a*) Officers holding permanent Commissions, and
(*b*) Officers holding short-service Commissions.

16. Officers holding permanent Commissions, on leaving Cranwell, are given the choice of the particular branch in which they wish to specialise. Those selecting the naval side will, in the first instance, do four years with the Fleet Air Arm. They may then serve with other branches, including courses of instruction, for a term of years in order to gain wider experience, but will ultimately return to the naval side again. This practice will ensure that there will always be a proportion of senior officers in the Royal Air Force with naval experience obtained in the Fleet Air Arm.

17. A proportion of the officers holding short-service Commissions for five years, after receiving their initial training of one year, are posted for duty with the Fleet Air Arm, and do the whole of their service with that branch. On completion of their five years' service they are posted to the Reserve for five years, when they are earmarked as reserves for the Fleet Air Arm in case of emergency.

18. The Air Ministry endeavour to retain the personnel with the Fleet Air Units for the normal term of service (*i.e.*, four years), but exceptional cases arise when it is impossible to allow them to remain for the whole of that period.

Training.

19. The whole of the primary training of flying personnel, whether their future work is to be on sea or on land, is carried out by the Air Ministry. No criticisms of it have been brought to our notice, and we do not propose to refer to it again. The later and more specialised training of the Fleet Air Arm is partly conducted in the carriers and partly on land at technical schools maintained by the Air Ministry at certain naval ports. Apart from their initial training, before they are embarked, the Air Detachments of Carriers use their establishments for training, when not required by the Naval Commander-in-Chief for exercise. It will be seen therefore that as things are at present these establishments are used entirely for the purposes of the Fleet Air Arm or for shore-based aircraft co-operating with the Navy. The Air Force training schools in the Portsmouth area work in close touch with the naval gunnery and torpedo schools. Behind the establishments devoted exclusively to training for the Fleet Air Arm are a series of technical centres maintained for the Air Force as a whole, the resources of which are available for the Fleet Air Arm.

Design and Research Work.

20. New design, experimental and research work to meet naval requirements are carried out by the Air Ministry after consultation with the Admiralty. The Admiralty, after consultation between the two Departments, put forward their suggestions as to the types of machines which they require, and the Air Ministry design such machines, as far as is practicable from the point of view of aerodynamics and similar technical consideration. The Air Ministry draw up the specification of the type required, and call for designs from the trade, and the Admiralty are kept fully informed of the progress made during the whole process and after the trials have been carried out.

21. The Navy are not directly represented on the design, experimental or research branches at the Air Ministry, but the Air Ministry utilise the services of a certain number of ex-naval officers, who have now transferred to the Air Force, in these departments, and also receive reports as to defects in design and performance of machines from Air Force officers serving with Fleet Air Units, thus ensuring that due weight is attached to the user's point of view in the modification and improvement of design.

22. There is also in existence an Inter-Departmental Committee called 'The Joint Technical Committee on Aviation Arrangements in His Majesty's Ships,' which consists of representatives of the Admiralty and the Air Ministry. Its functions are to examine the technical problems

which may arise with regard to the suitability of the types of aircraft intended for embarkation on board carriers, and the modifications and improvements which can be effected on board the carriers themselves. It is clear from the evidence that new uses of aircraft involving possibly new types can be proposed either by the Admiralty or the Air Ministry, and that the latter give full consideration to all suggestions which may be put forward by the Navy.

The Supply of Matériel.

23. At present the Air Ministry are solely responsible for the supply of all *matériel* connected with aircraft. An establishment of the machines required for the Fleet Air Arm has been agreed upon between the Admiralty and the Air Ministry, and the latter is responsible for maintaining that establishment at full strength with its correct proportion of machines in reserve, the necessary spare parts and the various shore organisations required for its maintenance. This establishment is at present limited by the number of Aircraft Carriers in commission in the Navy, but the Air Ministry have made the necessary arrangements in their programme of construction to increase the number of machines up to the number which will be required when the two new Aircraft Carriers are completed in 1923-24.

PART III

Recommendations

24. We have endeavoured to explain, in the first place, the general nature of the objection felt by the Air Department to any important change in the existing system and by the Admiralty to its maintenance. We have also given a full account of the system as it has been developed up to the present moment. We now propose to turn to the future and make certain suggestions which, as we hope, may go far to reconcile the contending parties and to make material improvements in the existing system.

25. Our task is materially lightened by the fact that, in the course of discussions before your Sub-Committee, a larger measure of agreement was found possible on certain aspects of the problem than had at one time seemed likely.

26. In the first place, it appeared that on some important points the practice, though not the written law, of the Air Department was already in conformity with the wishes of the Admiralty. For example, the Admiralty are very desirous that the Group Captains in contact with the Navy should be officers with experience of the Fleet Air Arm.

1st Recommendation.

Broadly speaking, the Air Ministry take the same view, and we recommend that it be made part of the settled practice of the Air Ministry to consult with the Admiralty before such appointments are made. We think the same principle should be extended to the Air Officer Commanding the Coast Defence Area and the Senior Air Officer on board a carrier.

27. The Admiralty desire that the Air Ministry shall provide all the *matériel* which they demand. This, we believe, has always been the intention of the Air Ministry.

2nd Recommendation.

We think it should be formally laid down, and that, in the event of the Air Ministry for one reason or another not carrying into effect the wishes of the Admiralty, there shall be an appeal by either party to the Committee of Imperial Defence.

28. We have no reason to believe that there is any fundamental difference of opinion between the two Services with regard to certain questions of discipline, such, for example, as those which might arise when a member of the Air Force leaves the ship, where he is under the command, for every purpose, of the Captain, and goes for training to the aerodrome, which is under the control of an officer of the Air Force.

3rd Recommendation.

But though there are no differences of principle, there are certain obscurities which it would be desirable to clear up, and we therefore recommend that the two Departments should, in collaboration, draw up a code of regulations which would make clear the relations between them at all points where they come in contact.

29. In all the cases that we have so far enumerated there may have been misunderstanding, but there has not been at any time, so far as we are aware, any difference of principle.

30. We now turn to a class of cases where, as a result of discussion before your Sub-Committee, the Air Department have shown a disposition to meet Admiralty views. The first of these relates to the framing of the Naval Estimates. The Admiralty desire that in these Estimates should be included the cost of the Fleet Air Arm.

4th Recommendation.

The Air Department are of opinion that technically this course would be inconvenient, but have no objection to it in principle. Your Sub-Committee are of opinion that this question should be settled by the Treasury in consultation with the two Fighting Services.

31. The Admiralty require that the Air Force on board ship should be completely under the orders of its Captain. Your Sub-Committee are of opinion that, inasmuch as all the airmen on board ship are under the Naval Discipline Act, this, in theory, is already provided for, but undoubtedly in the minds of those chiefly concerned some obscurity hangs over the subject. This, we think, should be explicitly cleared up.

5th Recommendation.

The position of a member of the Air Force when on board ship does not differ in law, and should not differ in practice, from the position of, say, a Marine.

6th Recommendation.

32. In order fully to carry out the policy laid down in the preceding paragraph, we recommend, and have reason to believe that the Air Ministry will accept the principle, that all reports on officers of the Fleet Air Arm, whether confidential or otherwise, should be signed by the Captain of the ship and passed through the Naval Commander-in-Chief to the Air Officer Commanding the Coastal Area.

7th Recommendation.

33. A further point which should be dealt with under this heading is one on which the Admiralty have expressed considerable anxiety and on which we have reason to believe that the Air Force are prepared to meet their wishes. The Admiralty are apprehensive lest, in time of war or other emergency, the Air Ministry might withdraw from the Navy units of the Fleet Air Arm, and use them for other purposes. In respect of this we recommend that it should be definitely laid down that the personnel, *matériel* and reserves of the Fleet Air Arm should not be withdrawn by the Air Ministry without either the consent of the Admiralty or a decision of the Cabinet.

34. Before concluding this class of questions we must take note of a complaint made by the Admiralty that on board the carrier there is a good deal of duplication of effort between the purely Naval Service and the Air Force.

8th Recommendation.

We think this subject should be looked into by the two Departments concerned, and we do not doubt that arrangements can be made by which all overlapping can be effectively avoided.

35. We now come to the last class of questions with which we have to deal, which are at once the most difficult and the most important. We are strongly of opinion, and we have every hope that the two Services share our view, that, since in war the Services may have to co-operate, it is vital that in time of peace they should form an accurate estimate of each others needs and capacities.

9th Recommendation.

We therefore recommend :

(i.) That Naval officers should be appointed to the Air Staff.

(ii.) That Air Force officers should be appointed to the Naval War Staff.

(iii.) That some means should be devised by which the wealth of technical knowledge at the disposal of the Admiralty should be utilised in the technical departments of the Air Ministry, preferably by reinforcement of the staff of these departments by naval technicians.

36. These recommendations refer exclusively to the headquarters of the two Services ; but it is perhaps even more important that the junior ranks of the two Services, who will in time occupy responsible positions, should have a considerable sprinkling of persons familiar with the needs and capacities of the other Service.

37. The Air Force look to a system of naval seconding for carrying out half this policy, and we cannot believe that the Admiralty would be averse to having members of the Air Force on board the carriers. Unfortunately, seconding from the Navy to the Air Force, as at present understood, must be deemed to have been hitherto a failure, and we find it difficult to believe that, if the present system remains unchanged and unexplained, any great improvement is likely to occur. We cannot be surprised that a young officer who has just joined the Navy is reluctant to abandon, even temporarily, the department under which he expected to serve and to exchange

it for one which is essentially different. The feeling is natural, and cannot be ignored.

It must, however, be pointed out that if the word 'seconding' is thus used this is not what any naval officer under the present scheme is expected to do. In ordinary practice, when we say that an officer is 'seconded' to another Service we no doubt intend to express the idea that for the duties which he had to perform and for the authorities he had to obey in the Service which he has temporarily left, will be substituted new duties and new authorities. But this does not really represent the facts in the case of so-called 'seconding' from the Navy into the Air Force. The duties of the seconded officer, though carried out in the air, remain, nevertheless, Naval duties, and the Captain whom he has to obey continues to be a Naval Captain.

10th Recommendation.

In order to make this situation perfectly clear, we recommend that no seconded Naval officer shall be asked to perform non-naval air duties, except with the consent of the Admiralty. We believe the Air Force are prepared to accept this principle.

38. If this recommendation be carried into effect, the most important change involved in the operation of what is (somewhat inaccurately) called 'seconding' is the change from the Naval to a Flying uniform—a change which can hardly be said to touch the essence of the situation.

11th Recommendation.

We suggest, however, in order to meet the sentimental, though not on that account unimportant, objection, that the uniform of a Naval flying man who, except for his period of training, is to all intents and purposes still under the Admiralty, should be distinguished from the flying men under the Air Force by some differentiating badge or mark. This would be the outward and visible sign that he still remains a member of the Service which he originally joined. It would be a clear indication that what he proposes to do is to add accomplishments in Naval flying to the other accomplishments which his brother officers are cultivating. If such a plan were found practicable, the objection felt by the Admiralty to the introduction of what they deem an alien element into the domestic life of the ship should be largely mitigated.

39. It might well be that, if this scheme succeeded, the number of officers seconded from the Navy to the Air Force would exceed the 30 per cent. of the total contemplated by the Air Service.

12th Recommendation.

We see no reason why 30 per cent. should be regarded as the maximum, and we should propose to leave it to the Admiralty to determine what the proportion should be, subject to the proviso that not less than 30 per cent. of Air Force officers, whether regular or short service, should serve on board the carriers.

40. There is another point of great practical importance on which something must be said. It is agreed that the work of spotting for naval gunnery is one which should be undertaken in all cases by Naval officers, but such investigations as we have been able to make convince us that naval spotting and fleet reconnaissance cannot be sharply divided, and

that the officer entrusted with the one may inevitably find himself called upon to perform the other also. We have been unable to discover any objection to this change, which, on the face of it, seems obviously reasonable.

13th Recommendation.

We therefore recommend that fleet reconnaissance, as well as naval spotting, should be entrusted to Naval officers seconded or otherwise.

41. If anybody will take the trouble to read the evidence given before us they will, we believe, be impressed by the number of problems for which a solution has been found by the two Departments or suggested in this Report. We earnestly trust that no merely technical difficulties will be allowed to stand in the way of a settlement which in the public interest is most urgently required.

(Signed) BALFOUR (*Chairman*).
PEEL.
WEIR.

M. P. A. HANKEY (*Secretary*).
2, WHITEHALL GARDENS,
July 21, 1923.

II.—THE CO-ORDINATION OF THE DEFENCE FORCES.

The following decisions have been taken by His Majesty's Government on the recommendations of the National and Imperial Defence Committee :

(1) It is undesirable and impracticable to supersede the Ministerial heads of the three Fighting Services by making them subordinates of a Minister of Defence ; the alternative plan for an amalgamation of the three Service Departments is equally impracticable.

(2) On the other hand, the existing system of co-ordination by the Committee of Imperial Defence is not sufficient to secure full initiative and responsibility for defence as a whole and requires to be defined and strengthened.

(3) Under the existing system the Committee of Imperial Defence, an advisory and consultative body, inquires into and makes recommendations in regard to the issues of defence policy and organisation which are brought before it. The power of initiative lies with the Government Departments and with the Prime Minister.

(4) This system, though invaluable up to a point, does not make any authority, except the Prime Minister, who can only devote a small part of his time and attention to defence questions, directly responsible for the initiation of a consistent line of policy directing the common action of the three or any two of the three Services, taking account of the reactions of the three Services upon one another.

(5) While, therefore, the existing system of departmental initiative will continue, the responsibility for the wider initiative referred

to above in Paragraph (4) will also rest with the Chairman of the Committee of Imperial Defence acting under the general direction of the Committee of Imperial Defence and with the assistance of the three Chiefs of Staff.

(6) In accordance with the terms of the Treasury Minute of May 4, 1904, constituting the Committee of Imperial Defence in its present form, the Committee of Imperial Defence will continue to consist of the Prime Minister, as President, with such other members as, having regard to the nature of the subject to be discussed, he may from time to time summon to assist him. In pursuance of a decision by the Prime Minister, the Committee places on record that the following should be members :

The Chairman (Deputy to the Prime Minister),
The Secretary of State for War,
The Secretary of State for Air,
The First Lord of the Admiralty,
The Chancellor of the Exchequer, or the Financial Secretary,
The Secretary of State for Foreign Affairs,
The Secretary of State for the Colonies,
The Secretary of State for India,
The Chiefs of Staff of the three Fighting Services,
The Permanent Secretary of the Treasury as head of the Civil Service.

In addition to these, other British or Dominion Ministers of the Crown and other officials, or persons having special qualifications, will be summoned as members by the President according to the nature of the business.

(7) The functions of the Chairman of the Committee of Imperial Defence will be :

(i.) To preside over the Committee of Imperial Defence in the absence of the Prime Minister.

(ii.) To report to the Prime Minister (when he himself has not presided) and to the Cabinet the recommendations of the Committee of Imperial Defence.

(iii.) In matters of detail, to interpret the decisions of the Prime Minister and the Cabinet thereupon to the Departments concerned.

(iv.) Assisted by the three Chiefs of Staff, as laid down in paragraph (5) above, to keep the defence situation as a whole constantly under review so as to ensure that defence preparations and plans and the expenditure thereupon, are co-ordinated and framed to meet policy, that full information as to the changing naval, military and air situation may always be available to the Committee of Imperial Defence and that resolutions as to the requisite action thereupon may be submitted for its consideration.

(8) In addition to the functions of the Chiefs of Staff as advisers on questions of sea, land or air policy respectively, to their own Board or Council, each of the three Chiefs of Staff will have an individual and collective responsibility for advising on defence policy as a whole, the three constituting, as it were, a Super-Chief of a War Staff in Commission. In carrying out this function they will meet together for the discussion of questions which affect their joint responsibilities.

(9) Questions relating to co-ordination of expenditure may be entertained by the Committee of Imperial Defence when referred to it by the Cabinet. The Committee (subject to any directions by the Cabinet) will consider such questions in the light of the general defence policy of the Government and of the strategical plans drawn up to give effect to that policy in time of war.

(10) The Secretariat of the Committee of Imperial Defence will continue to act as liaison officers between the Chairman of the Committee and the Service Departments. The staff of the Committee will be strengthened by the addition of an Assistant Secretary to be nominated by the Prime Minister on the recommendation of the Secretary of State for Air, whose status will be identical with that of the three existing Assistant Secretaries nominated by the Prime Minister on the recommendation of the Secretary of State for War, the Secretary of State for India and the First Lord of the Admiralty.

(11) The Standing Defence Sub-Committee is suppressed and its past proceedings will be merged into those of the Committee of Imperial Defence.

APPENDIX II.

REPORT

OF THE

SUB-COMMITTEE OF THE COMMITTEE OF IMPERIAL DEFENCE

ON

NATIONAL AND IMPERIAL DEFENCE.

Cmd. 2029. 1924.

APPENDIX II

SUB-COMMITTEE APPOINTED BY THE PRIME MINISTER TO INQUIRE INTO THE QUESTION OF NATIONAL AND IMPERIAL DEFENCE.

Terms of Reference

The Prime Minister desires that a Sub-Committee of the Committee of Imperial Defence, composed as follows:

The Marquess of Salisbury (*in the Chair*),
The Chancellor of the Exchequer,
The Secretary of State for Foreign Affairs,
The Secretary of State for the Colonies,
The Secretary of State for War,
The Secretary of State for India,
The Secretary of State for Air,
The First Lord of the Admiralty,
Lord Balfour,
Lord Weir,
Sir Maurice Hankey (*Secretary*),

should meet to enquire into the co-operation and correlation between the Navy, Army and Air Force from the point of view of National and Imperial Defence generally, including the question of establishing some co-ordinating authority, whether by a Ministry of Defence or otherwise, and, in particular, to deal with:

(*a*) The relations of the Navy and Air Force, as regards the control of Fleet air work.
(*b*) The corresponding relation between the Army and Air Force.
(*c*) The standard to be aimed at for defining the strength of the Air Force for purposes of Home and Imperial Defence.

(Initialled) A. B. L.

2, Whitehall Gardens, S.W.1.
March 9, 1923.

Part I.—INTRODUCTORY

An examination of the Terms of Reference to the Sub-Committee, which are given on the preceding page, reveals that the inquiry covers the following separate but closely connected groups of questions:—

(1) The co-operation and correlation between the three Services from the point of view of National and Imperial Defence (*see* Part II.).

(2) Their co-ordination, whether by means of a Ministry of Defence or otherwise (*see* Part III.).
(3) The relations of the Navy and Air Force as regards the control of Fleet air work (*see* Part IV.).
(4) The corresponding relations of the Army and Air Force (*see* Part V.).
(5) The standard to be aimed at for defining the strength of the Air Force for purposes of Home and Imperial Defence (*see* Part VI.).

The whole of the above questions were dealt with, though not in the same order, by the Sub-Committee itself, which is hereafter referred to as the Committee or the Main Committee, with the one exception of item (3), which was remitted to a special Sub-Committee composed of:

Lord Balfour,
Lord Peel, and
Lord Weir.

2. Between March 15 and October 31, the Main Committee has held 19 meetings, in addition to the 12 meetings of the special Sub-Committee, making a grand total of 31 meetings. The evidence of the Chiefs of Staff of the Fighting Services has been heard at great length on all the subjects discussed. The special Sub-Committee heard 16 witnesses. No less than 67 Memoranda were furnished to the Main Committee and 19 to the Sub-Committee, a total of 86 Memoranda. These include Memoranda on the question of a Ministry of Defence by the following authorities outside the Government service (*see* Part III.):

Lord Midleton,
Lord Haldane,
Major-General Seely,
Sir Eric Geddes,
Field-Marshal Sir William Robertson,
Lieut.-General Sir Aylmer Hunter-Weston,
Major-General Sir J. H. Davidson,
Major-General Sir Frederick Sykes.

In addition, the Committee derived useful information from a report called for by their special Sub-Committee from Lieutenant-General Sir Charles Harington, General Officer Commanding-in-Chief, Allied Forces in Occupation in Turkey.

On June 18, the Chairman, accompanied by the Secretary of State for India and Lord Weir, received a deputation from the Parliamentary Air Committee. The deputation attended primarily with the object of presenting views on the question of the relations of the Navy and Air Force, but touched on many of the related questions which were before the Committee.

PART II.—CO-OPERATION AND CORRELATION BETWEEN THE THREE SERVICES FROM THE POINT OF VIEW OF NATIONAL AND IMPERIAL DEFENCE

3. At the outset of the inquiry the Committee agreed that the term 'National and Imperial Defence' could properly be defined as 'Defence

of Territory and Defence of Communications.' In order to establish the principles of co-operation and correlation between the three Services from the point of view of National and Imperial Defence, it was found necessary to make a careful examination of the responsibilities of the three Services in this matter.

PRE-WAR RESPONSIBILITIES OF THE FIGHTING SERVICES

4. The main responsibilities of the Navy, as they were regarded before the war, were set forth in a Memorandum by the Committee of Imperial Defence on the Principles of Imperial Defence in 1910 in the following terms :

> 'The maintenance of sea supremacy has been assumed as the basis of the system of Imperial Defence against attack from over the sea. This is the determining factor in shaping the whole defensive policy of the Empire, and is fully recognised by the Admiralty, who have accepted the responsibility of protecting all British territory abroad against organised invasion from the sea.'

4. The Army was responsible for dealing not with organised invasion from the sea, which was a purely naval responsibility, but with such raiding forces as might elude the Fleet. This responsibility included the general military defence of the United Kingdom, as well as the provision of fixed defences and garrisons at Naval bases, Imperial coaling stations and defended ports at home and abroad (other than those in the Dominions and India). The Army was also responsible for the provision of reinforcements for India in certain eventualities, for the defence of certain land frontiers, and for the maintenance of an expeditionary force to meet the military needs of the Empire, wherever they might arise. This expeditionary force formed the nucleus on which were built up the huge armies employed in the Great War.

6. The Royal Air Force before the war was regarded as purely ancillary to the older Services, of which it formed a part. It emerged from the Great War as a separate Service under a separate Ministry. It is essential, therefore, to the co-ordination of Imperial Defence that the responsibilities of the new Service should be very clearly defined and correlated with those of the sister Services.

THE POST-WAR RESPONSIBILITIES OF THE FIGHTING SERVICES

7. Considerable progress had been made before the commencement of the present inquiry in the direction of defining the post-war responsibilities of the three fighting Services. The inquiry by Mr. Bonar Law's Sub-Committee on the Capital Ship (March, 1921), which heard much evidence as to the potentialities of the air arm at sea, had resulted in the retention of the capital ship as the basis of our sea power. The scope of the inquiry, however, did not extend to a definition of the respective responsibilities of the Navy and Air Force in regard to operations at sea.

8. The Committee of Imperial Defence had also approved an arrangement made between the War Office and Air Ministry in regard to the responsibility for anti-aircraft defence, under which the Air Force was to be responsible for the control of anti-aircraft defences, the War Office providing the necessary personnel and *matériel* on the ground.

9. Since the war, the Royal Air Force has been given the responsibility for the security of the mandated territory of Iraq and Palestine.

10. In addition, on March 16, 1922, the following principles in regard to the co-operation of the three Services were announced in the House of Commons :

'(i.) That the Air Force must be autonomous in matters of administration and education ;

'(ii.) That in case of defence against air raids, the Army and Navy must play a secondary *rôle ;*

'(iii.) That in the case of military operations by land or naval operations by sea, the Air Force should be in strict subordination to the General or Admiral in supreme command ; and lastly,

'(iv.) That in other cases (such as the protection of commerce, and attack on enemy harbours and inland towns) the relations between the Air Force and the other Services should be regarded rather as a matter of co-operation than that of the strict subordination which is necessary when aeroplanes are acting merely as auxiliaries to the other arms.' (Parliamentary Debates, March 16, 1922.)

11. In order to ascertain whether the above allocation of responsibilities between the three Services required further readjustment, the Committee felt it necessary to make investigations into the strategical basis of our system of National and Imperial Defence.

12. The most important result of this part of the inquiry was to confirm the vital need for a great increase in our air forces, which had been established in previous inquiries. It soon became clear, not only that the Air Force has an important part to play in the defence of our home territory against sea-borne attack as well as of maritime communications in waters adjacent to the British Islands, but that to provide protection against aerial attack a large aerial Home Defence Force was required. The question of the size of this force formed the subject of an interim report to the Cabinet, and is dealt with in Part VI. of this Report. The adoption by the Government of the Committee's recommendations on this subject added to the responsibilities of the Royal Air Force. In other respects, however, it did not bear upon the problem of co-operation and correlation between the three Services, which depends on those aspects of National and Imperial Defence where strategical or tactical co-operation between two or more Services is required. The Committee therefore inquired into these questions in considerable detail.

13. So far as the protection of territory and communications in the wider oceans is concerned, the question of a readjustment of responsibilities between the Service Departments did not arise. Although certain types of aeroplanes have a radius of action up to 500-600 miles, and others up to 300-400 miles, neither the British nor any other Air Service is yet equipped with aeroplanes whose normal effective radius of action exceeds about 200 miles, and beyond that distance only sea-borne aircraft have for the present to be considered. But it must be remembered that the types of aeroplanes now in service use continue steadily to be replaced by machines of greater power and wider radius of action.

Whatever the arrangements for the organisation and administration of the sea-borne air arm may be—a subject dealt with in Part IV.—there is no dispute that its operations must be controlled and directed by the same authority that controls and directs the other operations of the Fleet, namely, the Admiralty. The responsibility of the Admiralty, therefore, for the protection of territory and communications in great oceans was not challenged.

14. In the narrow seas, however, Imperial territory and communications are evidently liable to attack, not only by surface craft and submarines, but also by aircraft, in so far as they are within the radius of action of aircraft operating from foreign territory. The Committee therefore found it necessary to inquire how far the power of the Navy to protect territory and communications in the narrow seas is affected by modern developments of naval and aerial warfare.

15. So far as territory is concerned, it is satisfactory to be able to record that the three General Staffs are agreed that in existing conditions the liability of the United Kingdom to sea-borne invasion as compared with the years preceding the war is negligible.

16. Another matter on which some measure of agreement was found to exist between the Naval Staff and the Air Staff is in regard to the increased risks to communications in those portions of the narrow seas which are exposed to attack by aircraft operating from shore bases, such as the English Channel and the Mediterranean. Both Staffs admit that the advent of aircraft has increased the danger to communications in such waters, though they differ as to the extent of this danger.

17. Apart from this, when the Committee came to examine the question of the protection of maritime communications in the narrow seas, they did not find the same measure of agreement between the Staffs which they had met with in regard to the protection of the territory of the Mother Country. On the contrary, this part of the inquiry revealed wide differences of professional opinion between the Naval Staff and the Air Staff, both in matters of principle and detail, on such questions as the power of a fleet to operate within effective striking range of hostile aircraft, the effectiveness of attacks on a fleet by aircraft and the power of a fleet to defend itself against such attacks, the defence of naval bases against aircraft attack and the protection of certain portions of our trade routes. The more closely the inquiry was pressed in matters of detail the wider these differences appeared.

In the course of this inquiry frequent appeals were made to the experience of the war in support of both sides of the various controversies, and the Committee obtained independent historical evidence from the Historical Section. The conditions of the late war, however, are not necessarily applicable to future wars, particularly in view of the potentialities of development in aircraft.

In the light of present knowledge the Committee did not feel competent to form an opinion on the difficult technical questions on which the General Staffs take different views. They felt that, apart from the provision of an adequate Air Defence Force, which is dealt with in Part VI. of this Report, the most useful service they could render at the moment was to devise machinery for securing the smooth co-operation of the three Services, notwithstanding the differences of professional opinion, and for gradually building up a doctrine common to the three Services.

This part of the terms of reference is dealt with in Part III. of this Report. While the Committee do not put forward any suggestion to change the existing division of responsibility between the three Services as described in paragraph 10, they recommend that the responsibilities of the Navy and the Air Force in regard to the protection of communications in the narrow seas should form the subject of further investigation. They also consider that experiment in respect of the problems of air attack and defence at sea should be given due weight in Admiralty and Air Ministry programmes, in order to secure on the basis of practical experience the fullest measure of unity of professional opinion.

18. Before leaving the question of the co-operation and correlation between the three Services, the Committee desire to draw attention to the views of Lieutenant-General Sir Charles Harington on the subject, which were formed as the result of his experience in command of the Allied Forces of Occupation at Constantinople. These views illustrate and confirm the soundness of the principles approved by the Cabinet in March, 1922 (Paragraph 10). General Harington suggests 'that the principle should be laid down that the "predominant partner" co-ordinates and the other Services should conform in exactly the same loyal way in which Admirals de Robeck and Brock—both officers senior to me—have helped me.' While the Committee are not prepared to recommend without further technical examination that this principle should be adopted in the settlement of all military problems in which more than one Service is concerned, they consider that it is worthy of the most serious consideration and should be examined by the new Committee of Chiefs of Staff referred to later in this Report (Paragraph 36).

CONCLUSIONS

19. The conclusions of the Committee in regard to this part of the Report may be summed up as follows:

(*a*) While the Committee do not put forward any suggestion to change the existing division of responsibility between the three Services as described in Paragraph 10, they recommend that the responsibilities of the Navy and the Air Force in regard to the protection of communications in the narrow seas should form the subject of further investigation. The Committee further recommend that experiment in respect of the problems of air attack and defence at sea should be given due weight in Admiralty and Air Ministry programmes, in order to secure on the basis of practical experience the fullest measure of unity of professional opinion.

(*b*) The principal need, as regards co-operation and correlation, is closer co-ordination, which is dealt with in Part III. of this Report.

(*c*) The principle that in all belligerent operations in which more than one Service is concerned one of the three Services should be selected as a 'predominant partner' to co-ordinate the other Services should be examined by the Committee of Chiefs of Staff.

(*d*) While the menace of attack from the air has greatly increased and necessitates a strong Home Defence Air Force as proposed in Part VI. of this Report, the three Staffs are agreed that in existing conditions the liability of the country to sea-borne invasion has considerably diminished as compared with pre-war standards.

Part III.—CO-ORDINATION

A Ministry of Defence

20. Since the war the most widely-discussed proposal for overcoming our defects in co-ordination is a Ministry of Defence, which has been put forward repeatedly both in Parliament and in the Press. In 1922 Sir Eric Geddes' Committee on National Expenditure recommended 'the creation of a Co-ordinating Authority or a Ministry of Defence responsible for seeing that each force plays its part and is allotted appropriate responsibility for carrying out various functions.' A Cabinet Committee, which reviewed the Geddes Report, endorsed the above recommendation, but, while admitting the creation of a Ministry of Defence may be the ultimate solution of the problem, did 'not consider that the present time is appropriate for the fusion of the administration of the three Services under one Minister.' They recommended instead 'that the Committee of Imperial Defence should be in constant session all the year round in order to consider and advise on matters of policy affecting the three Fighting Services.' On March 21, 1922, the late Secretary of State for the Colonies (Mr. Churchill), during a debate in the House of Commons, made an important speech in favour of a Ministry of Defence as the ultimate solution of our problems of co-ordination, though he admitted that the time for accomplishing this had not arrived. The gist is contained in the final passages:

> 'No solution of a harmonious or symmetrical character will be achieved in the co-ordination of the Services except through the agency of a Ministry of Defence, but it is not possible to create such a body at the present time, nor will it be possible for a considerable time. In the interim the only steps which are open to us are to create machinery for pooling the administrative functions of the three arms and to create a common staff brain, from whose exertions in the future the responsible advice given to the Cabinet of the day in regard to matters of defence must and can only effectively originate.'

21. The interim steps referred to by Mr. Churchill were the appointment of a Committee, under Sir Alfred Mond, whose place was later taken by Lord Weir, to consider the amalgamation of the common Services of the Navy, Army and Air Force, and, later, of a Sub-Committee of the Committee of Imperial Defence under the Minister of Education, on the question of establishing a joint Staff College for the three Services. Lord Weir's Committee reported that 'the amalgamation of the common Services would only be practicable if it formed part of a comprehensive scheme of reorganisation which provided for the establishment of a Ministry to control a defence force in which the identity of the Navy, Army and Air Force had been merged.' They recommended, however, a complete scheme for co-ordinating the common Services, which has since

been adopted by the Cabinet and is being put in operation. The report of Mr. Wood's Committee, proposing a scheme for the formation of a Joint Staff College, has been circulated to this Committee and is now before the Committee of Imperial Defence.

22. In view of the uncertainty as to what is meant by the term 'Ministry of Defence,' the Committee thought it desirable to approach some of those who have advocated this solution of the problem of co-ordination, as well as other outside authorities, in order to ascertain their views on the subject. Among the outside experts who were good enough to give the Committee the benefit of their experience on this subject, there were only two advocates of a Ministry of Defence, and each of them contemplated its formation by different methods. Sir Eric Geddes proposed to achieve co-ordination of the Services by the creation of a single Secretary of State for 'Warfare' or 'Defence,' with responsibility for the three Services. Under the Secretary of State he would put the Admiralty, War Office and Air Ministry, each of which would be under a Sub-Minister. These Sub-Ministers should not be eligible for Cabinet rank, as that would make for departmental competition between them. Among themselves they would be equal in importance and would bear the same relation to the Secretary of State as a Parliamentary Under Secretary holds to his Ministerial chief to-day. The Sub-Ministers would preside respectively at the Board of Admiralty, Army Council and Air Council, which would remain more or less as at present. The Secretary of State would have a very small office, containing a Statistical Accountant and a Council consisting of the Sub-Ministers of the Navy, Army and Air Force, with two members each from the Board of Admiralty, Army Council and Air Council. The Secretary of State would have to obtain the endorsement of the Committee of Imperial Defence 'before his estimates and his provisions were taken to the Cabinet.' Sir Eric Geddes added proposals for organisation after the outbreak of war, which it is unnecessary to enter into here.

23. Major-General Sir Frederick Sykes advocated a Ministry of Defence by means of a form of fusion or amalgamation of the existing Service departments. He considered that 'the real solution lies in definite, unified, supreme control by a Defence Ministry, with the Prime Minister as independent Chairman, and a joint Staff which would really think out defence as a whole.' Failing this policy of perfection, General Sykes would 'support every measure which will pave the way for such control.' He would achieve this by 'the real strengthening of the mandate and constitution of the Committee of Imperial Defence. In his view, the Committee should frame estimates for defence for the three Services, and a special section should be formed for the specific purpose of jointly framing and supervising major schemes and measures of defence. As many Services as possible should be unified. The personnel of the regular arms should be trained from boyhood. There should be a joint boy-mechanic training, a joint Cadet College, a joint Staff course, etc. The question of the amalgamation of similar Services should be reopened.'

24. The proposals of the remaining outside authorities had many points in common. They all laid stress on the need for co-ordination. They all proposed that this co-ordination should be effected through the Committee of Imperial Defence. They nearly all admitted, either directly or by implication, that the work of directing the Committee was too heavy

for the Prime Minister to undertake single-handed, and that he should have the assistance of a Minister, as Vice-Chairman, who could give most of his time to this task. Several laid stress on the importance of securing the co-operation of the Dominions in the Committee of Imperial Defence.

25. In detail, the proposals varied considerably. Lord Midleton and Lord Haldane did not indicate any particular developments of the existing Committee of Imperial Defence, the case for which was summed up by the latter in the following terms:

> 'The Committee of Imperial Defence is an organisation that has nothing quite resembling it in any other country. The reason is that no other nation resembles the British Empire, with its Island centre for a number of countries, some of which are self-governing, and all of which are united by unwritten and elastic obligations. We have evolved this Committee to meet Dominion as well as Home necessities, and to meet the former it is far better adapted than any special Ministry of Defence could be. . . . But a still more distinctive feature of the existing organisation is that it has been evolved to meet a situation where sea power comes first and where the other two Services are, in some measure, merely its adjuncts, however great and important. That is why the scope of the Committee must be sufficiently catholic to admit of the co-operation within it of distinguished experts at the head of very different Services.'

The above extract was specifically endorsed by the Chief of the Imperial General Staff.

26. Major-General Seely, Field-Marshal Sir William Robertson, Major-General Sir J. H. Davidson, M.P., and Lieut.-General Sir Aylmer Hunter-Weston, M.P., all advocated definite extensions of the Committee of Imperial Defence organisation, which resembled one another in principle while differing in detail.

27. General Seely's principal proposal was 'that a Minister must be appointed, under whatever title be deemed expedient, whose sole duty it will be to secure the co-operation of the three Services, reporting fully to the Prime Minister and the Cabinet. I suggest that he should be styled "Minister of Defence and Vice-President of the Committee of Imperial Defence."' General Seely strongly advocated the retention of the civilian heads of the three Service Departments in their existing status as members of the Cabinet. 'Great Government Offices,' he writes, 'cannot possibly be controlled efficiently by any man without Cabinet rank. I am sure that anyone who has had experience of Great Government Offices would agree with this view.'

28. Field-Marshal Sir William Robertson's proposals are summarised in his Memorandum as follows:

> '(*a*) Neither a Ministry of Defence nor a combined Imperial General Staff will provide, or help to provide, the co-ordinating authority we require.
>
> '(*b*) Controlling authority, in its true meaning, must be vested in the supreme executive power, the Cabinet, and it cannot be placed elsewhere.
>
> '(*c*) Experience has shown that we cannot conduct a great war through the medium of a Cabinet of twenty or more Ministers,

and that the duty is best assigned to a small body of Ministers having no other duties to perform. As this organisation is not feasible in peace time, its place should be taken by a Council of Imperial Defence, which will form a nucleus for war.

'(*d*) In order to furnish this Council with the professional assistance needed, there should be, working under it, a Technical Committee, charged with the investigation of all operative and administrative questions, and with presenting them, with recommendations thereon, to the Council for consideration and approval.

'(*e*) In time of war a Minister of Supply and a Minister of Man-power should be appointed and a War Cabinet should be formed. The latter, assisted and advised by the three Chiefs of Staff, would take over the duties in (*c*) and (*d*).

'(*f*) Every effort should be made to enlist the co-operation of the Dominions, both as to State policy and war preparations.'

29. The following details of Sir William Robertson's plan may be added. The Committee would be composed very much as heretofore. In addition:

'The Prime Minister would, of course, be President. It would be the duty of the three Chiefs of Staff to advise the Council on professional matters, the advice to be taken or left as Ministers may think best, but to be heard. These officers should, as was the War Cabinet system, attend the Council in an advisory capacity and not as members. I believe this procedure to be the best for both parties. Following the pre-war constitution of the Committee of Imperial Defence, a senior officer from each of the three Services should be included in the Council as members. Their experience would enable them to give valuable help both to Ministers and the Chiefs of Staff, and they would be specially useful to the former in cases where the latter might feel compelled to differ from each other in regard to professional matters upon which they were called upon to advise. If these three officers (unemployed) are not included in the Council, Ministers will have to decide for themselves professional questions about which their knowledge must necessarily be imperfect, and the soundness of their decisions will, therefore, remain largely a matter of chance.'

30. Major-General Sir J. H. Davidson, M.P., referred the Committee to an article published by him in the *Army Quarterly* of January, 1921, in which he advocated a plan presented to the Government by the Parliamentary Army Committee in June, 1920, the essentials of which are contained in the following extract from a Minute addressed by the Committee to the Prime Minister (Mr. Lloyd George):

'4. It is possible that the most practicable scheme under present conditions would be to create immediately a Standing Joint Defence Sub-Committee of the Committee of Imperial Defence, formed of the First Sea Lord, the C.I.G.S., and the C.A.S., or officers appointed or deputed by them, together with representatives of the Self-Governing Dominions, of India, and of other Departments concerned.

'5. Whatever be the Advisory Body formed, the members of the Army Committee in the House of Commons are of opinion that it is essential:

'(i.) That it should meet regularly and frequently.

'(ii.) That it should have a specially selected and permanent Secretariat to assist in its work, and to record its proceedings and conclusions.

'(iii.) That the Chairman of this Sub-Committee should be a Minister not in charge of one of the Great Departments of State, except on those occasions when the Prime Minister is himself present.

'6. Among the duties of the Standing Joint Defence Sub-Committee of the C.I.D., should be included the following:

'To examine:

'(*a*) The Imperial Organisation for Defence.

'(*b*) The Estimates, in draft, with a view to ensuring due economy and efficiency.

'(*c*) Our Imperial responsibilities from the point of view of Defence.

'(*d*) The effect of scientific progress and inventions.

'(*e*) The problems of Strategy and Logistics.

'(*f*) The proposals of the League of Nations.'

31. Lieut.-General Sir Aylmer Hunter-Weston, M.P., was a signatory of the Minute of the Parliamentary Army Committee referred to above, and his Memorandum gives a carefully thought-out scheme for applying the principles propounded therein. He recommended the retention of the present Committee of Imperial Defence with the title of Council of Imperial Defence, under the Chairmanship of the Prime Minister, with the Lord President of the Council as Vice-Chairman. In addition, he proposed the establishment of a Board of Defence, composed of the Ministerial and professional heads of the three Fighting Services, with a representative of the Treasury, meeting under the Chairmanship of the Vice-Chairman of the Council of Defence, who might with advantage be called Minister of Defence. The Chairman should have power of initiative in all matters of strategy, policy, and finance affecting more than one Service. The duties proposed for this 'Board of Defence' under its Chairman are similar in character to those suggested by the Parliamentary Army Committee for the Standing Joint Defence Sub-Committee, but worked out in greater detail.

32. An essential part of General Hunter-Weston's plan was that the Council of Imperial Defence should function actively and regularly, and that the Dominions and India should be represented at its meetings. He advocated that, as no one man can fulfil the functions of Chief of the General War Staff of our Defence Forces, the office should be put in commission and carried out by the professional heads of the three Services sitting together in Committee.

33. In the view of the Committee the proposals for a Minister of Defence are effectually disposed of in the following passage from Lord Haldane's Memorandum:

'In the way of the institution of a general Minister of Defence there are obvious difficulties. If established with anything like adequate power of control, such a Minister would be bound to interfere in administration, just as the First Lord of the Admiralty and the Secretary of State for War are bound to be ready to do so, by reason of their direct responsibility for it to Parliament. The Minister of Defence would, indeed, be looked to as responsible not only for efficiency, but for economy. He would therefore require a considerable and varied staff, whose duties would overlap and duplicate those of existing departmental staffs. What would be the relation of this new staff to the staffs under the three Ministers at present responsible to Parliament, and what would be the constitutional and practical relationship of the new Minister of Defence to the three older Ministers? The former would, I think, be in considerable danger of proving himself to be either too great or too little. He would be too little if the departmental staffs developed to their full inherent capacity and were working out general military policy in conference. In such a case the Prime Minister would be the only person possessed of authority sufficient to enable him to intervene effectively.

'With the Cabinet behind him, he is in a position to exercise influence as no Minister of Defence could.

'If, indeed, the Minister of Defence were to make himself, on the other hand, very powerful by equipping himself with an effective administrative organisation sufficient for direct control of the three Services, he might well become a rival of the Prime Minister himself. The difficulty does not exhaust itself here. The first Government that made such an appointment would probably make it with great care and with sufficient regard to necessary qualifications in the occupant of the position. But if a subsequent Government came in that were not deeply interested in defence, the temptation would be strong to give the office to an influential politician distinguished, perhaps, mainly for debating gifts.'

34. To these objections may be added the following, urged by Lord Midleton :

'It is surely beyond human power for one man to get his mind impregnated with the pros and cons of large changes in three totally distinct Services within the limited time for which Parliamentary Chiefs hold office. The fact that there have been eleven changes in the Office of Secretary of State for War in the last eleven years has been very prejudicial to the economy and possibly to the efficiency of the Army. First Lords of the Admiralty attach the greatest importance to their official tours for elucidating by contact with Naval officers not employed at the Admiralty the problems submitted to them. The overworked Minister of Defence would be quite unable to find time for such excursions.

'A further difficulty would be the Parliamentary one, since it is often necessary for the Minister in charge to give a pledge during a debate as to the course which his Department will take. Not infrequently it has happened that by far the most efficient Head of the Defence Committee would be a Member of the House of Lords. . . . If the supreme executive responsibility of all three Departments were to be massed in one Minister it would be imperative that he should sit in the House of Commons, and attendance in the House of Commons would add immeasurably to the already multifarious duties imposed on him.'

35. The closely-connected question of a combined Staff is disposed of equally effectively by Field-Marshal Sir William Robertson in the following terms :

'The formation of a combined Imperial General Staff, consisting of Military, Naval and Air Force officers, working under a Chief (a soldier, or sailor, or airman) responsible to the Government, or to a Minister of Defence, for working out plans of operations on land, on sea, and in the air, and, according to some, endowed with "financial and strategical powers," is even more fantastical as well as dreadfully mischievous. An important corner-stone in military organisation is that he who makes a plan ought to be responsible for its execution and stake his reputation upon it. Consequently, the Chief of this proposed combined Staff must draft and issue the orders of the Government to all the Generals and Admirals and Air Officers entrusted with the control of the armies, the fleets and the air forces. The confusion that would arise in the three War Departments and at the front, if any such ill-considered system as this were adopted, is quite inconceivable. Further, this Staff would directly interpose between the three Chiefs of Staff and the Cabinet, and there could be no more pernicious system than that.'

36. The Committee considered these criticisms to be overwhelming as against all proposals for setting up a Ministry of Defence or any Minister of Defence with authority overriding that of the Ministers at the head of the Service Departments, or a combined Staff. After careful consideration of the various proposals laid before them, after a full discussion with the Chiefs of Staff of the three Fighting Services, and after a close examination of the constitution and the present methods of work of the Committee of Imperial Defence, the Committee reached the following conclusions, which were adopted by the Government and presented to Parliament in August last (Cmd. 1938) :

'(1) It is undesirable and impracticable to supersede the Ministerial heads of the three Fighting Services by making them subordinates of a Minister of Defence ; the alternative plan for an amalgamation of the three Service Departments is equally impracticable.

'(2) On the other hand, the existing system of co-ordination by the Committee of Imperial Defence is not sufficient to secure full initiative and responsibility for defence as a whole and requires to be defined and strengthened.

'(3) Under the existing system the Committee of Imperial Defence, an advisory and consultative body, inquires into and makes recommendations in regard to the issues of defence policy and organisation which are brought before it. The power of initiative lies with the Government Departments and with the Prime Minister.

'(4) This system, though invaluable up to a point, does not make any authority, except the Prime Minister, who can only devote a small part of his time and attention to defence questions, directly responsible for the initiation of a consistent line of policy directing the common action of the three or any two of the three Services, taking account of the reactions of the three Services upon one another.

'(5) While, therefore, the existing system of departmental initiative will continue, the responsibility for the wider initiative referred to above in paragraph (4) will also rest with the Chairman of the Committee of Imperial Defence acting under the general direction of the Committee of Imperial Defence and with the assistance of the three Chiefs of Staff.

'(6) In accordance with the terms of the Treasury Minute of May 4, 1904, constituting the Committee of Imperial Defence in its present form, the Committee of Imperial Defence will continue to consist of the Prime Minister, as President, with such other members as, having regard to the nature of the subject to be discussed, he may from time to time summon to assist him. In pursuance of a decision by the Prime Minister, the Committee places on record that the following should be members :

'The Chairman (Deputy to the Prime Minister),
'The Secretary of State for War,
'The Secretary of State for Air,
'The First Lord of the Admiralty,
'The Chancellor of the Exchequer, or the Financial Secretary,
'The Secretary of State for Foreign Affairs,
'The Secretary of State for the Colonies,
'The Secretary of State for India,
'The Chiefs of Staff of the three Fighting Services,
'The Permanent Secretary to the Treasury as head of the Civil Service.

In addition to these, other British or Dominion Ministers of the Crown and other officials, or persons having special qualifications, will be summoned as members by the President according to the nature of the business.

'(7) The functions of the Chairman of the Committee of Imperial Defence will be :

'(i.) To preside over the Committee of Imperial Defence in the absence of the Prime Minister.

'(ii.) To report to the Prime Minister (when he himself has not presided) and to the Cabinet the recommendations of the Committee of Imperial Defence.

'(iii.) In matters of detail, to interpret the decisions of the Prime Minister and the Cabinet thereupon to the Departments concerned.

'(iv.) Assisted by the three Chiefs of Staff, as laid down in paragraph (5) above, to keep the defence situation as a whole constantly under review so as to ensure that defence preparations and plans and the expenditure thereupon are co-ordinated and framed to meet policy, that full information as to the changing naval, military and air situation may always be available to the Committee of Imperial Defence and that resolutions as to the requisite action thereupon may be submitted for its consideration.

'(8) In addition to the functions of the Chiefs of Staff as advisers on questions of sea, land or air policy respectively, to their own Board or Council, each of the three Chiefs of Staff will have an individual and collective responsibility for advising on defence policy as a whole, the three constituting, as it were, a Super-Chief of a War Staff in Commission. In carrying out this function they will meet together for the discussion of questions which affect their joint responsibilities.

'(9) Questions relating to co-ordination of expenditure may be entertained by the Committee of Imperial Defence when referred to it by the Cabinet. The Committee (subject to any directions by the Cabinet) will consider such questions in the light of the general defence policy of the Government, and of the strategical plans drawn up to give effect to that policy in time of war.

'(10) The Secretariat of the Committee of Imperial Defence will continue to act as liaison officers between the Chairman of the Committee and the Service Departments. The staff of the Committee will be strengthened by the addition of an Assistant Secretary to be nominated by the Prime Minister on the recommendation of the Secretary of State for Air, whose status will be identical with that of the three existing Assistant Secretaries nominated by the Prime Minister on the recommendation of the Secretary of State for War, the Secretary of State for India, and the First Lord of the Admiralty.

'(11) The Standing Defence Sub-Committee is suppressed, and its past proceedings will be merged into those of the Committee of Imperial Defence.'

The above recommendations are now in operation.

The Representation of the Dominions and India

37. There is one point in the above conclusions on which the Committee, in this their final Report, would like to add a few observations, namely, regarding the provision in Conclusion (6) for the extension of invitations to representatives of the Dominions to attend as members of the Committee. From the earliest days of the Committee of Imperial Defence, representatives of the Dominions have from time to time been invited to take part in its proceedings, and the Secretary of State for India has for many years attended its meetings.

38. The subject of Dominion representation was discussed at the Committee in May, 1911, when all the Dominions were represented and the following resolution was passed :

> 'That one or more representatives appointed by the respective Governments of the Dominions should be invited to attend meetings of the Committee of Imperial Defence, when questions of naval and military defence affecting the Overseas Dominions are under consideration.'

39. Before the war every effort was made to give effect to the above resolution by taking advantage of the presence in this country of representatives of the Dominions, to invite them, with the concurrence of their own Governments, to the meetings of the Committee and of its Sub-Committees. Since the war no such opportunities have offered, though in fact many of the meetings of the Imperial Conference, 1921, were analogous to meetings of the Committee of Imperial Defence; those present included the regular members of the Committee, the subjects were mainly those which naturally fall to the Committee, and the secretarial work was supplied by the Committee in co-operation with the Colonial Office, the Dominions and India.* Moreover, both before and since the war the Dominions and India have been furnished with many of the Reports of the Committee of Imperial Defence.

Part IV. THE RELATIONS OF THE NAVY AND THE AIR FORCE

40. The question of the relations of the Navy and the Air Force in regard to fleet air work had already been the subject of acute and prolonged controversy, and was recognised to require investigation of a detailed character. As already mentioned, therefore, this question was referred to a special Sub-Committee composed of:

Lord Balfour,
Lord Peel,
Lord Weir.

Their Report, together with the remarks of the main Committee thereon, were adopted by the Government, and presented to Parliament last August. (Cmd. 1938.) These Reports are annexed to this Report for convenience of reference.

Part V.—THE RELATIONS OF THE ARMY AND THE AIR FORCE

41. The question of the relations of the Army and the Air Force formed the subject of an Interim Report by the Chairman to the Cabinet, dated June 30, 1923, the effective portions of which are contained in the following extracts:

> 'The Secretary of State for War recently circulated to the Sub-Committee a Memorandum by the General Staff on this question. . . .
> 'The views of the General Staff are contained in Part I. of that Memorandum, and are summarised on p. 4 in the following terms:
>
> '"(a) Their units, which are an integral part of the Fleet and Air formations (including probably lighter-than-air formations) capable of co-operating with the Fleet on the high seas, to be under the Admiralty.

* *Note by the Chairman.*—This is equally true of the Imperial Conference, 1923. The question of defence was discussed by the plenary conference at great length, and in similar conditions to those mentioned in this Report. In addition, important discussions took place at the Admiralty and Air Ministry. (*See* the Official Summary of the proceedings of the Imperial Conference, Cmd. 1987.)

'" (*b*) The air units, which are an integral part of Army formations and Air formations required to co-operate with the Army (including Air Forces allotted to the general pool for war and to home defence), to be under the War Office.

'" (*c*) Civil aviation, research, experiment and supply to be under the Air Ministry, which, relieved of all responsibility for the employment of Air Forces in peace and war, could be much reduced.

'" Each of the three Departments (Naval, Military and Civil) would estimate for its own air requirements, the whole being co-ordinated by the Committee of Imperial Defence before presentation to Parliament."

'The view of the Sub-Committee was to the effect that the above distribution of responsibility was unsatisfactory, and that if the Air Forces of this country were to be developed to the utmost it was necessary to retain the Royal Air Force as a separate Service, and that progress would not be so great if the War Office proposals were adopted. The conclusion of the Sub-Committee was:

'" That they were unable to accept the views expressed by the General Staff in Part I. of their Paper." '

The Secretary of State for War, while strongly supporting the General Staff solution, expressed the readiness of the War Office to co-operate in furthering the scheme as accepted.

The Cabinet adopted the above Report on July 9 and confirmed the present arrangement under which the Royal Air Force is administered by the Air Ministry as a separate Department of State.

PART VI.—THE STRENGTH OF THE ROYAL AIR FORCE

42. The last item in the Terms of Reference is the standard to be aimed at for defining the strength of the Air Force for purposes of Home and Imperial Defence.

43. This question was dealt with in an Interim Report, dated June 12, 1923.

This Report was approved by the Cabinet on June 20, 1923, and the following announcement of the Government's policy based thereon was made in both Houses of Parliament on the same date:

'The Government have come to the following conclusions with reference to British Air power:

'In addition to meeting the essential Air power requirements of the Navy, Army, Indian and Overseas commitments, British Air power must include a Home Defence Air Force of sufficient strength adequately to protect us against Air attack by the strongest Air Force within striking distance of this country.

'It should be organised in part on a regular and permanent military basis, and in part on a territorial or reserve basis, but so arranged as to ensure that sufficient strength will be immediately available for purposes of defence. The fullest possible use to be made of civilian labour and facilities.

'In the first instance, the Home Defence Force should consist of 52 squadrons, to be created with as little delay as possible, and the Secretary of State for Air has been instructed forthwith to take the preliminary steps for carrying this decision into effect. The result of this proposal will be to add 34 squadrons to the authorised strength of the Royal Air Force. The details of the organisation will be arranged with a view to the possibility of subsequent expansion, but before any further development is put in hand the question should be re-examined in the light of the then Air strength of foreign Powers.

'In conformity with our obligation under the Covenant of the League of Nations, His Majesty's Government would gladly co-operate with other Governments in limiting the strength of air armaments on lines similar to the Treaty of Washington in the case of the Navy, and any such arrangements, it is needless to say, would govern the policy of air expansion set out in this statement.' (Parliamentary Debates, June 26, 1923.)

44. In order to arrive at the total standard of strength for the Royal Air Force, it is necessary to add to the forces for Home Defence the strength of:

Royal Air Force Units serving with the Navy.

The Air Forces required for co-operation with the Army.

The Air Forces to be maintained by the Air Ministry to fulfil their responsibilities in Iraq and Palestine.

45. It is not possible at the present time to make a recommendation as to the ultimate standard to be aimed at for any of the Services mentioned in paragraph 44. As regards the Royal Air Force Units serving with the Navy, the ultimate strength in war will be determined in the main by the number of first line aeroplanes which can be carried by the Fleet, that is to say, by carriers, battleships and light cruisers.

46. As regards the Air Forces for co-operation with the Army, the Secretary of State for Air has given an assurance to the Secretary of State for War that any requirements will be fully satisfied, subject to Treasury approval. The actual figures require further detailed discussion between the Chiefs of the General Staffs. This consultation was held up pending the decision on the relations of the Army and the Air Force. The Committee recommend that these discussions should be pushed forward as rapidly as possible by the Chiefs of Staff under the new procedure proposed in Part III.

47. Similarly, the ultimate establishment of the Air Forces abroad involves many uncertain factors, such as the future of Iraq and Palestine, and the number of machines eventually to be provided for Singapore.

48. The Committee are not concerned in the number of squadrons in India, which are paid for by India and are a matter of negotiation between the Government of India, the India Office and the Air Ministry. The subject, however, has recently been fully explored by the Committee on Indian Military Requirements.

49. In the above circumstances, the Committee can only record their recommendation that there must be sufficient air strength for the Navy,

the Army, the Overseas Garrisons and Home Defence. The squadrons and machines authorised up to April 1, 1924, exclusive of Home Defence, are as follows:

	Machines.
United Kingdom:	
(*a*) (i.) *Fleet Air Arm.*—For embarkation in carriers in all waters (13 flights)	78
(ii.) *Naval Co-operation.*—Flying Boats (1 flight)	5
(*b*) Army co-operation (2 squadrons)	24
(*c*) Reserve (3 squadrons)	36
Mediterranean:	
Seaplanes (1 flight)	6
(Aircraft for embarkation in carriers shown under United Kingdom above.)	
Egypt (3 squadrons)	34
Palestine and Transjordania (1½ squadrons)	16
Aden (1 flight)	4
Iraq (8 squadrons)	92
Total	295
India (6 squadrons)	72
Grand Total of machines	367

Note.—Army co-operation machines are not differentiated except in the United Kingdom.

Airships

50. A question which is closely related to the main inquiry is that of airships, which had been referred to the Committee of Imperial Defence before the appointment of this Committee. At the first meeting this question was remitted to an Inter-Departmental Committee composed of representatives of the Admiralty, Air Ministry and Treasury, which was set up on the initiative of the Air Ministry, and was asked to report direct to the Committee of Imperial Defence. The policy of the Government on this question was announced in Parliament on July 26 in the following terms:

'The question of the development of airships has recently been considered by the Committee of Imperial Defence and the Imperial Shipping Committee. The Committee of Imperial Defence attaches considerable strategic value to airships, whilst the Imperial Shipping Committee considers that it is by means of an airship service that the carriage of mails can most cheaply be expedited to the Far East and Australia.

'The Government have, therefore, decided to resume the development of airships, and to proceed, if possible, by means of a commercial service rather than by State operation.

'Proposals have been placed before them by the hon. and gallant Member for Uxbridge (Lieutenant-Commander Burney), under which a bi-weekly service of six large airships to India will eventually be set up. The Government have accepted the scheme in principle, subject to the details of the contract being satisfactorily settled by the Treasury. The House of Commons will have an opportunity of considering the scheme when the details have been provisionally agreed.

'The Dominions are being informed of this decision, and it is hoped to discuss the question at the Imperial Conference with a view to their co-operation in the scheme.

'The administration of the scheme in so far as it is a matter of commercial aviation will come under the Air Ministry.' *

Part VII.—SUMMARY OF CONCLUSIONS AND RECOMMENDATIONS

51. The conclusions and recommendations of this Report may be summarised as follows :

Co-operation and Correlation of the Services (paragraph 19).

(*a*) While the Committee do not put forward any suggestion to change the existing division of responsibility between the three Services as described in paragraph 10, they recommend that the responsibilities of the Navy and the Air Force in regard to the protection of communications in the narrow seas should form the subject of further investigation. The Committee further recommend that experiment in respect of the problems of air attack and defence at sea should be given due weight in Admiralty and Air Ministry programmes, in order to secure on the basis of practical experience the fullest measure of unity of professional opinion.

(*b*) The principal need as regards co-operation and correlation is closer co-ordination, which is dealt with in Part III. of this Report.

(*c*) The principle that in all belligerent operations in which more than one Service is concerned, one of the three Services should be selected as a 'predominant partner' to co-ordinate the other Services should be examined by the Committee of Chiefs of Staff.

(*d*) While the menace of attack from the air has greatly increased, and necessitates a strong Home Defence Air Force, as proposed in Part VI. of this Report, the three Staffs are agreed that in existing conditions the liability of the country to sea-borne invasion has considerably diminished as compared with pre-war standards.

* *See* Imperial Economic Conference, 1923. Summary of Conclusions, page 9 (Cmd. 1990); also Record of Proceedings and Documents, page 351 *et seq.* (Cmd. 2009).

Co-ordination of the Services (*paragraph* 36).

(*e*) It is undesirable and impracticable to supersede the Ministerial heads of the three Fighting Services by making them subordinates of a Minister of Defence; the alternative plan for an amalgamation of the three Service Departments is equally impracticable.

(*f*) On the other hand, the existing system of co-ordination by the Committee of Imperial Defence is not sufficient to secure full initiative and responsibility for defence as a whole, and requires to be defined and strengthened.

(*g*) Under the existing system the Committee of Imperial Defence, an advisory and consultative body, inquires into and makes recommendations in regard to the issues of defence policy and organisation which are brought before it. The power of initiative lies with the Government Departments and with the Prime Minister.

(*h*) This system, though invaluable up to a point, does not make any authority, except the Prime Minister, who can only devote a small part of his time and attention to defence questions, directly responsible for the initiation of a consistent line of policy directing the common action of the three or any two of the three Services, taking account of the reactions of the three Services upon one another.

(*i*) While, therefore, the existing system of departmental initiative will continue, the responsibility for the wider initiative referred to above in paragraph (*h*) will also rest with the Chairman of the Committee of Imperial Defence acting under the general direction of the Committee of Imperial Defence and with the assistance of the three Chiefs of Staff.

(*j*) In accordance with the terms of the Treasury Minute of May 4, 1904, constituting the Committee of Imperial Defence in its present form, the Committee of Imperial Defence will continue to consist of the Prime Minister, as President, with such other members as, having regard to the nature of the subject to be discussed, he may from time to time summon to assist him. In pursuance of a decision by the Prime Minister, the Committee places on record that the following should be members:

The Chairman (Deputy to the Prime Minister),
The Secretary of State for War,
The Secretary of State for Air,
The First Lord of the Admiralty,
The Chancellor of the Exchequer, or the Financial Secretary,
The Secretary of State for Foreign Affairs,
The Secretary of State for the Colonies,
The Secretary of State for India,
The Chiefs of Staff of the three Fighting Services,
The Permanent Secretary to the Treasury as head of the Civil Service.

In addition to these, other British or Dominion Ministers of the Crown and other officials, or persons having special qualifications, will be summoned as members by the President according to the nature of the business.

(*k*) The functions of the Chairman of the Committee of Imperial Defence will be :

(i.) To preside over the Committee of Imperial Defence in the absence of the Prime Minister.

(ii.) To report to the Prime Minister (when he himself has not presided) and to the Cabinet the recommendations of the Committee of Imperial Defence.

(iii.) In matters of detail, to interpret the decisions of the Prime Minister and the Cabinet thereupon to the Departments concerned.

(iv.) Assisted by the three Chiefs of Staff, as laid down in paragraph (*i*) above, to keep the defence situation as a whole constantly under review so as to ensure that defence preparations and plans and the expenditure thereupon are co-ordinated and framed to meet policy, that full information as to the changing naval, military and air situation may always be available to the Committee of Imperial Defence and that resolutions as to the requisite action thereupon may be submitted for its consideration.

(*l*) In addition to the functions of the Chiefs of Staff as advisers on questions of sea, land or air policy respectively, to their own Board or Council, each of the three Chiefs of Staff will have an individual and collective responsibility for advising on defence policy as a whole, the three constituting, as it were, a Super-Chief of a War Staff in Commission. In carrying out this function they will meet together for the discussion of questions which affect their joint responsibilities.

(*m*) Questions relating to co-ordination of expenditure may be entertained by the Committee of Imperial Defence when referred to it by the Cabinet. The Committee (subject to any directions by the Cabinet) will consider such questions in the light of the general defence policy of the Government, and of the strategical plans drawn up to give effect to that policy in time of war.

(*n*) The Secretariat of the Committee of Imperial Defence will continue to act as liaison officers between the Chairman of the Committee and the Service Departments. The staff of the Committee will be strengthened by the addition of an Assistant Secretary to be nominated by the Prime Minister on the recommendation of the Secretary of State for Air, whose status will be identical with that of the three existing Assistant Secretaries nominated by the Prime Minister on the recommendation of the Secretary of State for War, the Secretary of State for India and the First Lord of the Admiralty.

(*o*) The Standing Defence Sub-Committee is suppressed, and its past proceedings will be merged into those of the Committee of Imperial Defence.

The Relations of the Navy and the Air Force (*paragraph* 40).

(*p*) See Annex. (Appendix I.)

The Relations of the Army and the Air Force (*paragraph* 41).

(*q*) No change in the existing relations is recommended.

The Strength of the Royal Air Force (*paragraphs* 42 *to* 49).

(*r*) The standard of strength of the Royal Air Force is the sum of the following:

The Home Defence Air Force.
The Royal Air Force Units serving with the Navy.
The Air Forces required for co-operation with the Army.
The Air Forces maintained by the Royal Air Force to fulfil their responsibilities in Iraq and Palestine.

The figures, excluding those for the Home Defence Air Force, so far as they can at present be arrived at, are given in paragraph 49.

(*c*) The Air Forces required for co-operation with the Army should be worked out as soon as possible (paragraph 46).

(Signed) SALISBURY (*Chairman*).
NEVILLE CHAMBERLAIN.
CURZON OF KEDLESTON.
DEVONSHIRE.
DERBY.
PEEL.
SAMUEL HOARE.
L. S. AMERY.
BALFOUR.
WEIR.

M. P. A. HANKEY (*Secretary*).
2, WHITEHALL GARDENS, S.W. 1.
November 15, 1923.

APPENDIX III.

IMPERIAL CONFERENCE, 1923

DEFENCE RESOLUTIONS.

Cmd. 1987. 1923.

APPENDIX III

XII.—DEFENCE

The Conference gave special consideration to the question of Defence, and the manner in which co-operation and mutual assistance could best be effected after taking into account the political and geographical conditions of the various parts of the Empire.

The Lord President of the Council, as Chairman of the Committee of Imperial Defence, opened this part of the work of the Conference by a statement outlining the main problems of Defence as they exist to-day. He was followed by the First Lord of the Admiralty, the Secretary of State for War and the Secretary of State for Air, each of whom explained to the Conference the aspects of defence which concerned his special responsibilities.

In addition to these statements there was a full and frank interchange of views in which the standpoints of the various representatives and the circumstances of their countries were made clear. There were also discussions at the Admiralty and Air Ministry at which Naval and Air Defence were dealt with in greater detail. The points involved were explained by the Chiefs of the Naval and Air Staffs respectively and were further examined.

In connection with Naval Defence one matter of immediate interest came before the Conference, namely, the projected Empire Cruise of a squadron of modern warships. The First Lord of the Admiralty explained that the project was that two capital ships, the *Hood* and the *Repulse*, together with a small squadron of modern light cruisers, should visit South Africa, Singapore, Australia and New Zealand, and return by way of British Columbia, the Panama Canal and Eastern Canada. The light cruisers would accompany the battle cruisers as far as British Columbia, but would return to England by way of the west coast of South America and Cape Horn. The Dominion Prime Ministers expressed their appreciation of this proposal, and assured the Conference that the ships would be most heartily welcomed in their countries.

After the whole field of Defence had been surveyed, the Conference decided that it would be advisable to record in the following resolutions its conclusions on the chief matters which had been discussed :

'(1) The Conference affirms that it is necessary to provide for the adequate defence of the territories and trade of the several countries comprising the British Empire.

(2) In this connection the Conference expressly recognises that it is for the Parliaments of the several parts of the Empire, upon the recommendations of their respective Governments, to decide the nature and extent of any action which should be taken by them.

'(3) Subject to this provision, the Conference suggests the following as guiding principles :

(*a*) The primary responsibility of each portion of the Empire represented at the Conference for its own local defence.

(*b*) Adequate provision for safeguarding the maritime communications of the several parts of the Empire and the routes and waterways along and through which their armed forces and trade pass.

(*c*) The provision of Naval bases and facilities for repair and fuel so as to ensure the mobility of the fleets.

(*d*) The desirability of the maintenance of a minimum standard of Naval Strength, namely, equality with the Naval Strength of any foreign power in accordance with the provisions of the Washington Treaty on Limitation of Armament as approved by Great Britain, all the self-governing Dominions and India.

(*e*) The desirability of the development of the Air Forces in the several countries of the Empire upon such lines as will make it possible, by means of the adoption, as far as practicable, of a common system of organisation and training and the use of uniform manuals, patterns of arms, equipment, and stores (with the exception of the type of aircraft), for each part of the Empire as it may determined to co-operate with other parts with the least possible delay and the greatest efficiency.

'(4) In the application of these principles to the several parts of the Empire concerned the Conference takes note of :

(*a*) The deep interest of the Commonwealth of Australia, the Dominion of New Zealand, and India, in the provision of a Naval Base at Singapore, as essential for ensuring the mobility necessary to provide for the security of the territories and trade of the Empire in Eastern Waters.

(*b*) The necessity for the maintenance of safe passage along the great route to the East through the Mediterranean and the Red Sea.

(*c*) The necessity for the maintenance by Great Britain of a Home Defence Air Force of sufficient strength to give adequate protection against air attack by the strongest air force within striking distance of her shores.

'(5) The Conference, while deeply concerned for the paramount importance of providing for the safety and integrity of all parts of the Empire, earnestly desires, so far as is consistent with this consideration, the further limitation of armaments, and trusts that no opportunity may be lost to promote this object.'

PRINTED BY THE ANCHOR PRESS, LTD., TIPTREE, ESSEX, ENGLAND.

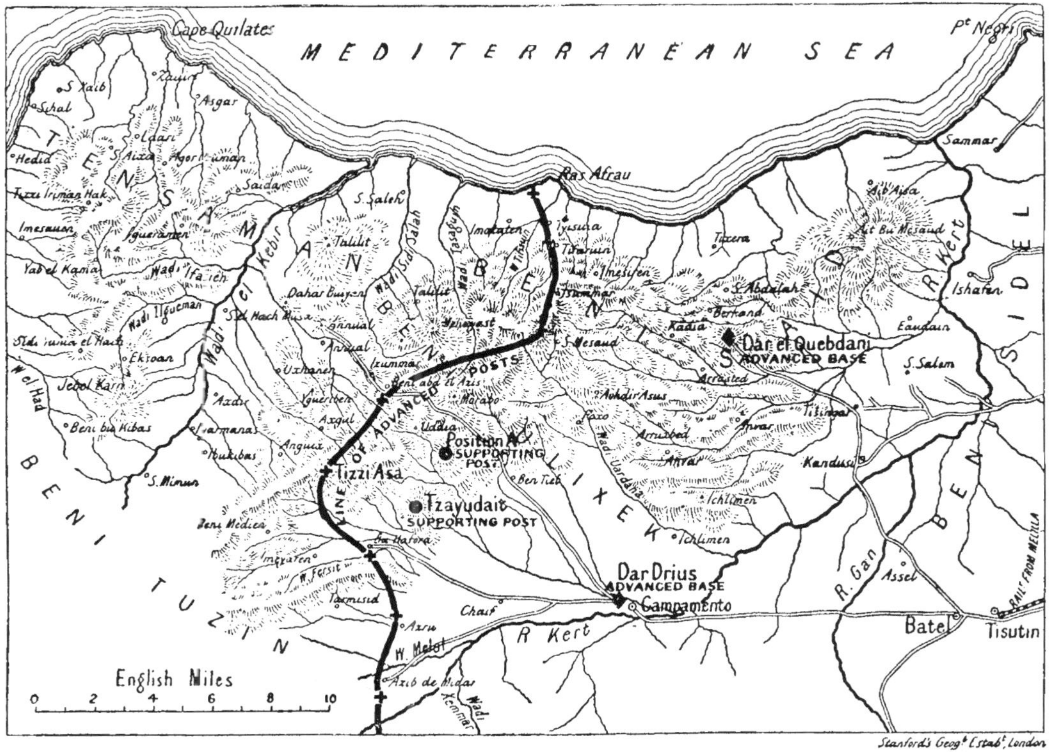

PART OF THE SPANISH ZONE IN MOROCCO.

p. 110.

Related Titles from Westphalia Press

The Limits of Moderation: Jimmy Carter and the Ironies of American Liberalism
by Leo P. Ribuffo

The Limits of Moderation: Jimmy Carter and the Ironies of American Liberalism is not a finished product. Yet, this book is a close and careful history of a short yet transformative period in American political history, when big changes were afoot. and continue to shape our world.

Bunker Diplomacy: An Arab-American in the U.S. Foreign Service
by Nabeel Khoury

After twenty-five years in the Foreign Service, Dr. Nabeel A. Khoury retired from the U.S. Department of State in 2013 with the rank of Minister Counselor. In his last overseas posting, Khoury served as deputy chief of mission at the U.S. embassy in Yemen (2004-2007).

Energy Law and Policy in a Climate-Constrained World
by Victor Byers Flatt, Alfonso López de la Osa Escribano, Aubin Nzaou-Kongo

This book presents reflections on concepts, foreign policy, regional and international cooperation, and the specific role the state is to play when it comes to such thing as energy law and policy.

The Zelensky Method
by Grant Farred

Locating Russian's war within a global context, The Zelensky Method is unsparing in its critique of those nations, who have refused to condemn Russia's invasion and are doing everything they can to prevent economic sanctions from being imposed on the Kremlin.

The Lord of the Desert: A Study of the Papers of the British Officer John B. Glubb in Jordan and Iraq
by Dr. Sa'ad Abudayeh

John Bajot Glubb, a British engineer officer, was sent to Iraq in 1920 to resolve the problems which erupted after the Iraqi revolt. He remained in the area for ten years, working with the Bedouins. In 1930, he moved to Jordan for twenty-six successful years. He invented what Dr. Abudayeh calls the Diplomacy of Desert.

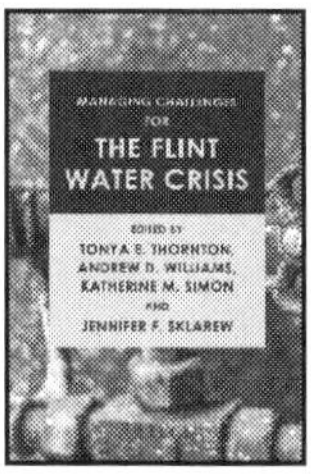

Managing Challenges for the Flint Water Crisis
Edited by Toyna E. Thornton, Andrew D. Williams, Katherine M. Simon, Jennifer F. Sklarew

This edited volume examines several public management and intergovernmental failures, with particular attention on social, political, and financial impacts. Understanding disaster meaning, even causality, is essential to the problem-solving process.

Resistance: Reflections on Survival, Hope and Love
Poetry by William Morris, Photography by Jackie Malden

Resistance is a book of poems with photographs or a book of photographs with poems depending on your perspective. The book is comprised of three sections titled respectively: On Survival, On Hope, and On Love.

Abortion and Informed Common Sense
by Max J. Skidmore

The controversy over a woman's "right to choose," as opposed to the numerous "rights" that abortion opponents decide should be assumed to exist for "unborn children," has always struck me as incomplete. Two missing elements of the argument seems obvious, yet they remain almost completely overlooked.

Geopolitics of Outer Space: Global Security and Development
by Ilayda Aydin

A desire for increased security and rapid development is driving nation-states to engage in an intensifying competition for the unique assets of space. This book analyses the Chinese-American space discourse from the lenses of international relations theory, history and political psychology to explore these questions.

The Athenian Year Primer: Attic Time-Reckoning and the Julian Calendar
by Christopher Planeaux

The ability to translate ancient Athenian calendar references into precise Julian-Gregorian dates will not only assist Ancient Historians and Classicists to date numerous historical events with much greater accuracy but also aid epigraphists in the restorations of numerous Attic inscriptions.

The Politics of Fiscal Responsibility: A Comparative Perspective
by Tonya E. Thornton and F. Stevens Redburn

Fiscal policy challenges following the Great Recession forced members of the Organisation for Economic Co-operation and Development (OECD) to implement a set of economic policies to manage public debt.

China & Europe: The Turning Point
by David Baverez

In creating five fictitious conversations between Xi Jinping and five European experts, David Baverez, who lives and works in Hong Kong, offers up a totally new vision of the relationship between China and Europe.

Issues in Maritime Cyber Security
Edited by Dr. Joe DiRenzo III, Dr. Nicole K. Drumhiller, and Dr. Fred S. Roberts

The complexity of making MTS safe from cyber attack is daunting and the need for all stakeholders in both government (at all levels) and private industry to be involved in cyber security is more significant than ever as the use of the MTS continues to grow.

Freemasonry, Heir to the Enlightenment
by Cécile Révauger

Modern Freemasonry may have mythical roots in Solomon's time but is really the heir to the Enlightenment. Ever since the early eighteenth century freemasons have endeavored to convey the values of the Enlightenment in the cultural, political and religious fields, in Europe, the American colonies and the emerging United States.

Growing Inequality: Bridging Complex Systems, Population Health, and Health Disparities
Editors: George A. Kaplan, Ana V. Diez Roux, Carl P. Simon, and Sandro Galea

Why is America's health is poorer than the health of other wealthy countries and why health inequities persist despite our efforts? In this book, researchers report on groundbreaking insights to simulate how these determinants come together to produce levels of population health and disparities and test new solutions.

Contests of Initiative: Countering China's Gray Zone Strategy in the East and South China Seas
by Dr. Raymond Kuo

China is engaged in a widespread assertion of sovereignty in the South and East China Seas. It employs a "gray zone" strategy: using coercive but sub-conventional military power to drive off challengers and prevent escalation, while simultaneously seizing territory and asserting maritime control.

Brought to Light: The Mysterious George Washington Masonic Cave
by Jason Williams, MD

The George Washington Masonic Cave near Charles Town, West Virginia, contains a signature carving of George Washington dated 1748. Although this inscription appears authentic, it has yet to be verified by historical accounts or scientific inquiry.

Frontline Diplomacy: A Memoir of a Foreign Service Officer in the Middle East
by William A. Rugh

In short vignettes, this book describes how American diplomats working in the Middle East dealt with a variety of challenges over the last decades of the 20th century. Each of the vignettes concludes with an insight about diplomatic practice derived from the experience.

westphaliapress.org

Policy Studies Organization

The Policy Studies Organization (PSO) is a publisher of academic journals and book series, sponsor of conferences, and producer of programs.

Policy Studies Organization publishes dozens of journals on a range of topics, such as European Policy Analysis, Journal of Elder Studies, Indian Politics & Polity, Journal of Critical Infrastructure Policy, and Popular Culture Review.

Additionally, Policy Studies Organization hosts numerous conferences. These conferences include the Middle East Dialogue, Space Education and Strategic Applications Conference, International Criminology Conference, Dupont Summit on Science, Technology and Environmental Policy, World Conference on Fraternalism, Freemasonry and History, and the Internet Policy & Politics Conference.

For more information on these projects, access videos of past events, and upcoming events, please visit us at:

www.ipsonet.org

Made in United States
Orlando, FL
27 February 2023

30484505R00183